ASPECTS OF ADHESION 5

aspects of ADHESION

5

proceedings of the conferences held at The City University, EC1 on 5 and 6 April 1967 and 9 and 10 April 1968

edited by **Professor D. J. Alner** M.Sc., Ph.D., F.R.I.C.

UNIVERSITY OF LONDON PRESS LTD

SBN 340 09639 X

University of London Press Ltd
St Paul's House, Warwick Lane, London EC4

Printed in Great Britain by
C. Tinling and Co. Ltd, London and Prescot

Preface

Aspects of Adhesion—5 contains the papers read at the conferences held in 1967 and 1968. As in previous volumes, the mixture of theory and technology has been maintained in an attempt to provide for the general needs of people interested or involved in the science and technology of adhesion and adhesives. For simplicity each topic discussed appears in the order in which it was presented at the meetings, the first nine papers referring to the 1967 conference and the last eight papers to the 1968 conference.

Quite apart from the need to develop satisfactory methods of testing adhesively bonded joints which are capable of interpretation and correlation with theory, a very great technical need exists for a simple, reliable and foolproof device capable of assessing the integrity and quality of adhesively bonded surfaces. A review of the state of knowledge on the testing of adhesive bonds was the central theme of the 1967 meeting. The concept of surface energy has been widely invoked to provide, at least in part, some explanation of the phenomenon of adhesion; although significant progress has been made, the situation is still far from satisfactory, and as a central feature the 1968 conference considers some aspects of the part surface energy plays in adhesion.

I wish to acknowledge all who made the two conferences so successful, contributors of papers and delegates, and in particular my colleague Mr K. W. Allen who was responsible for much of the detailed work of organisation and for the smooth running of the meetings. I express my thanks to the various publishing houses for their permission to reproduce illustrative material, and to the University of London Press Ltd for their continuing interest and support in publishing these papers.

D. J. ALNER

Contents

Proceedings of the 1968 Conference

Contributors

K. W. Allen, M.Sc., F.R.I.C.
The City University, London
R. S. Sharpe, B.Sc., F.Inst.P., F.I.M.
A.E.R.E., Harwell
L. Greenwood, B.Sc.
E.R.D.E., Waltham Abbey
P. E. Gallant, B.Sc., A.R.C.S., A.R.I.C.
Imperial Metal Industries (Kynoch) Limited
C. S. Swaffer
Imperial Metal Industries (Kynoch) Limited
W. Thompson, C. Eng., C.G.I.A., A.F.R.Ae.S.
A.I.D. Laboratories, Harefield
A. W. Henderson, B.Sc., Ph.D.
The Plessey Company Limited
E. F. Hall, F.B.S.I.
The Shoe and Allied Trades Research Association
N. C. MacDonald, B.Sc., A.R.I.C.
A.W.R.E., Aldermaston
D. F. Gentle, A.R.I.C.
Ford Motor Company Limited

Professor D. J. Alner, M.Sc., Ph.D., F.R.I.C.
The City University, London

C. A. Dahlquist, B.Ch.E., Ph.D.
Minnesota Mining and Manufacturing Company, U.S.A.

C. Weaver, B.Sc., F.Inst.P., M.I.E.E., M.I.E.E.E.
The University of Strathclyde, Glasgow

B. W. Malpass, B.Sc., Ph.D.
Passfield Research Laboratories, Metal Containers Limited

K. Bright, B.Sc., Ph.D.
Formica International Limited

T. L. Wright, A.C.T. (Birm.), A.P.I.
The Gelatine and Glue Research Association

Professor R. C. Pitkethly, B.Sc., Ph.D., F.R.I.C.
B.P. Research Centre, Sunbury, and The City University, London

E. W. Garnish, B.Sc., A.R.C.S.
CIBA (A.R.L.) Limited

C. G. Haskins, B.Sc.
CIBA (A.R.L.) Limited

M. J. Richardson, B.Sc., Ph.D.
National Physical Laboratory, Teddington

Theories of Adhesion Surveyed

K. W. ALLEN, M.Sc., F.R.I.C.
THE CITY UNIVERSITY, LONDON

Introductory

Over the last few years there has been a great deal published on the topic of adhesion. These publications range from the frankly sales promotion leaflets to abstruse theoretical treatises on the more recondite aspects; but fairly wide reviews seem to be somewhat rare. Particularly on the more theoretical aspects, a number of fairly well defined and recognised theories of adhesion have been advanced. Each one is supported by a number of protagonists and published in some particular journal but there is, so far as I can discover, no wide survey of the whole field. This may at least in part be due to the fact that there is no single journal wherein work on adhesion is generally published. It is this survey and review of the whole field which is attempted here.

The theories may be classified both by date and by geography and four principal ones can be distinguished:

(1) Mechanical. This is the oldest and apparently has its roots firmly in this country.

(2) Adsorption. This is perhaps the most highly regarded both here and in the United States. It underlies virtually the whole of Eley's book, and has as its protagonists e.g. Wake, Schonhorn.

(3) Electrostatic. This is essentially a Russian theory due initially to Deryaguin.

(4) Diffusion. Again a Russian theory propounded by Voyutskii and developed by Vasenin and others.

Both of these last two have suffered from the fact that most of the significant work has been published in Russian and is not readily accessible.

Mechanical theory

The earliest work on the technology of adhesion was very largely directed towards the needs of the aircraft industry and particularly joints involving wood as adherend. One of the distinctions made, particularly by McBain and Hopkins,[1] distinguished between specific adhesion and mechanical adhesion. They suggested that, at least in the case of wood and other materials of a porous nature, a prime factor in adhesion was the mechanical embedding of solidified glue in the pores and irregularities of the bonded surfaces. Thus the adhesive has to key into the porosities and irregularities of the materials.

In a situation where this sort of mechanism is of major significance two factors would seem to be immediately apparent:

(1) The greater the surface irregularity and porosity the greater the joint strength.

(2) Provided that the adherend is stronger than the adhesive, then the joint strength will be proportional to the film strength of the adhesive.

It is common knowledge in carpentry that a glued joint involving the end-grain of wood is always most liable to be a weak joint; and moreover the more open-grained the wood the weaker the joint. This is shown in a more precise and quantitative way by the following data:

TABLE I

Maplewood samples, bonded with urea-formaldehyde resin at 5 p.s.i. gluing pressure and tested in shear.[2]

Surface	*Shear strength* p.s.i.	
Planed	3 120	
Sanded	2 360	
Sawn	2 690	Increasing roughness ↓
Combed	2 400	

Clearly here, the smoother the surface and the less the opportunity for mechanical interlocking the stronger the joint!

The result of sanding and combing is to raise the fibres in the surface and to remove the softer material between them. These raised fibres are to a greater or lesser extent damaged and if they are firmly incorporated into the bond they can be removed from the bulk of the wood by quite small forces. Further the surface cavities are not usually involved since the adhesive does not enter them because of the air trapped in them.

In the case of end-grain the natural lumen of the grain will only be effectively involved in the bonding if they are reasonably clean-cut and not bent over and damaged as they are by normal saw-cuts; otherwise they will not be penetrated by the adhesive.

Now while this all tends to disprove a mechanical process of adhesion there are other instances where the evidence confirms its importance.

When some experimental precast concrete components for air raid shelters were being made during the war they were cast in moulds lined with Portland Stone and a quite unexpected almost complete adhesion was found between the castings and their moulds. This eventually was traced to the similarity in size between the particles of the cement and the natural interstices of the stone. While fuller analysis showed that there was also some chemical interaction and bonding, nevertheless it was demonstrated that approximately 90 per cent of the bond was mechanical.[3]

Also in the case of the bonding of rubber to textiles it has been shown conclusively by Borroff and Wake[4] to be essentially a case of mechanical adhesion, the strength of the bond depending upon the number of fibres which are embedded in the rubber.

Thus while it is not difficult to show that some of the examples which at first sight appear to be explained by a simple mechanical effect are in fact far more complex, yet there are examples where it may be conclusively demonstrated that mechanical effects are of prime importance.

Adsorption theory

The adsorption theory of adhesion, as its name implies, depends upon surface forces for its explanation of the phenomena observed. Essentially it regards adhesion as one particular property of a phase interface wherein polar molecules or groupings will be orientated in an ordered way.

Most of the thinking in this field has been done in connection with quite simple systems which are not directly concerned with adhesion. All the work of the surface chemists provides a background of knowledge and understanding of what happens at the surface between two different phases, but it is as well to be a little cautious in the extrapolation of their results to include those of real interest in adhesion.

The forces which are initially involved in this mechanism are the van der Waals' forces—the orientation, the induction and the dispersion effects which were all discussed fairly extensively by myself in the first Conference on Adhesion (1963)[5] and Dr. Ottewill in the second Conference (1964).[6] While so far there seems to be little or no correlation between the calculated and observed values for strengths on this basis, experiments on the attractive forces between pairs of quartz and mica plates do seem to have yielded very promising results, particularly over the range of 5 000Å to 10 000Å which is about the size of a coiled polymer molecule.

If a molecule, which is large enough to contain a polar group and a non-polar part as separate and distinct entities, approaches an interface at which the dielectric constant changes, it will orientate itself. It will move so that its non-polar part will be in the medium of lower dielectric constant and its polar part in the medium of high dielectric constant. This is the process which occurs in adsorption and must really take place if an adsorption theory is valid. While it has been argued that this is impractical on grounds that space would prevent it occurring in a solid adhesive or that thermal agitation would destroy it, nevertheless it must be remembered that in practice adhesives are always used in a form which is easily deformable.

If the adsorption theory were correct and were the whole explanation, then necessarily there should be a direct correlation between the energy of adsorption and the strength of bonding. While in a very general way there is a tendency for this to be observed, it is not sufficient to form the basis for a precise and quantitative theory.

Also adsorption requires, by standard physicochemical principles, that the active molecule shall be preferentially deposited from the solution on to the adsorbing surface. Again this cannot be regularly demonstrated for adhesives.

The problem which arises in trying to review the ramifications of this work are that it extends so far and in so many directions. It includes all the energetics and thermodynamics of surfaces, the flow and rheological properties of the adhesives and the kinetics of wetting.

Fortunately most of it has been published in the English language and is reasonably easily accessible, and I would remind you of Eley's book[7] which is essentially an exposition of this theory and its applications.

Perhaps it will suffice if I mention two particular points which have recently been well illustrated and expounded by Schonhorn.[8]

If two surfaces could ever be brought sufficiently close together that the molecules could really approach and interact then van der Waals' forces would become operative and they would stick together, and moreover they would be held together far more strongly than the real strengths of either of the separate materials. So to achieve any really useful joint we must arrange to get these materials closer together over a really useful area. Thus their surface roughness must be made to correspond, which essentially means that one of them must be a fluid. However fluidity alone is not sufficient, for a highly viscous liquid with a large contact angle will not make effective contact as it will bridge and entrap air in most of the surface depressions. Moreover stress concentrations due to the large contact angle may be significant. On the other hand a mobile liquid with small or zero contact angle which will spread easily will flow into the crevices and achieve true contact with little opportunity for stress concentration. Thus 'wettability' is of prime importance.

In all real adhesive problems the surfaces which are involved are not pure chemical surfaces any more than they are geometric planes. There is always a considerable degree of surface contamination and this frequently constitutes a weak boundary layer. It is quite well known that a number of polymers (e.g. polyethylene) can only be bonded effectively after a surface pretreatment e.g. exposure to corona or glow discharge, flaming, or chemical attack. It was held that these treatments provided an easily wetted surface by superficial oxidation but this alone is not sufficient and it was then suggested that it was due to an increase of the strength of this surface layer by cross-linking. This view is upheld and confirmed by recent work on the adhesion of polyethylene to steel surfaces and also of printing inks on polyethylene by Bright.[9]

Electrostatic theory

The electrostatic theory of adhesion arises from two initial observations obtained from peel testing—the first that the work involved in separating a film from a surface was of the order of 10^4 to 10^6 erg/cm^2 and the second that the observed adhesion of a film depends upon the speed at which it is peeled. Neither of these seem consistent with theories based on the usual surface tension effects, nor did the phenomena of crackling noises and flashes of light which may be observed when a film is peeled off rapidly in the dark.

Deryaguin and Krotova[10] took the well known phenomenon of an electrical double layer formed at the junction of two materials and applied this to a study of adhesion. The idea was that at any boundary an electrical double layer is produced and the consequent coulombic attraction might account for the adhesion and resistance to separation.

The theory begins from Paschen's law of electric discharge through a gas which states that the potential, V, giving rise to a spark discharge is proportional to the quantity of gas between the electrodes, that is to the product of the spark length, l, and the gas pressure, p:

$$\text{or } V \propto lp$$

If now the adherend plus the adhesive are considered to be one plate of a condenser and the film to be the other plate then, as they are separated, discharge may occur.

For a parallel plate condenser it can be shown[11] that

$$\text{Energy} \quad W = \tfrac{1}{2}\, eV = \tfrac{1}{2}\, cV^2$$

$$\text{and} \quad c = \frac{k\,a}{4\,\pi\,d}$$

$$\text{hence} \quad V^2 = \frac{8\,\pi\,W\,d}{k\,a} = \frac{8\,\pi\,W}{k\,a\,p} \times d.p$$

where W is energy
e is total charge
V is potential difference
c is capacity
k is dielectric constant
d is separation of plates
a is area of plates

Now some value can be assumed for W, and then for various values of p, work isobars can be calculated relating the potential difference and the product dp, which is essentially the amount of gas between the plates. These isobars can be plotted on the same diagram as the Paschen law curve. The intersection of an isobar with the Paschen law line gives the discharge voltage for a particular gas pressure as shown diagramatically in Fig. 1. Furthermore the energy of a condenser may also be expressed in the form

$$W = \frac{2\,\pi\,\sigma^2\,d}{k} \qquad \text{where } \sigma \text{ is surface charge density.}$$

Eliminating d between the two expressions for W gives:

$$W = \tfrac{1}{2}\, V\sigma$$

Deryaguin and his school made the assumption that the energy W could be equated with the work of adhesion and quote the following data for the adhesion of polyvinyl chloride to glass in an atmosphere of argon:

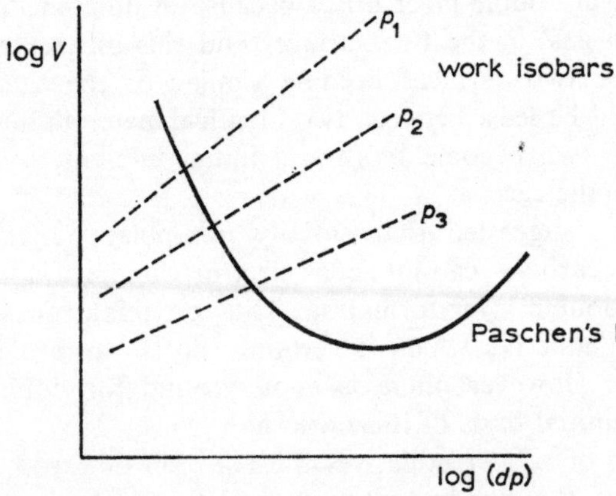

Fig. 1. Diagrammatic representation of the intersection of the Paschen's Law curve with a series of work isobars for several pressures

TABLE 2

Pressure (mm Hg)	Work of adhesion (erg/cm)	Condenser discharge energy (from graphs)	Charge density (cgs units)
760	1·8 × 10	1·8 × 10	2·9 × 10
175	3·2 × 10	3·1 × 10	3·0 × 10
100	4·0 × 10	4·1 × 10	2·8 × 10
50	5·0 × 10	5·1 × 10	2·8 × 10

The constant value of the charge density would seem to suggest that the theory is essentially sound, although a careful review suggests that the arguments are at least partially circular.

The crucial point however, is that this electrostatic theory requires a variation of the work of adhesion with the nature and pressure of the gas. The Russians claim to have observed these variations provided that rigorous drying is effected and stripping is carried out at fairly high speeds. Skinner, Savage and Rutyler[12] in the United States were unable to detect surface charges of the magnitude which Deryaguin et al. thought were involved.

This theory has been considerably criticised by exponents of alternative theories. Voyutskii, an advocate of the diffusion theory, suggests;[13]

(1) The electrical double layer arises because of different 'pressures of the electron gas' in the two surfaces and this difference will be minimal as the two surfaces become similar or the same. Thus adhesion should be least between two identical materials and should increase as the two become more and more different. In fact the exact reverse is the case.

(2) It has been suggested that entirely non-polar polymers, e.g. saturated hydrocarbons, cannot act as electron donors and so cannot contribute to double layer formation. Thus on this argument they should be non-adhesive which is certainly not in accord with the observed facts. However, there is good ground for doubting the validity of the initial basis of this criticism.

(3) In the case of rubber adhesives the addition of carbon blacks, which increases the conductivity, ought to reduce the adhesive properties by allowing charge leakage. Also the small amount of sulphur present in rubber and the reaction of vulcanization would not be expected to alter the contact potential and so the adhesive properties should not be affected. In both these cases the observed effect is just the reverse of these theoretical predictions; carbon-loaded rubbers retain their adhesive properties while vulcanized rubbers lose them.

Again I personally have considerable misgivings about the assumption upon which these predictions are based and certainly would anticipate considerable changes in surface potentials consequent on any alteration as radical as vulcanization of rubber.

Schonhorn,[8] on the other hand, who mainly supports the adsorption theory, criticised Deryaguin's work on quite different grounds, claiming that the peel tests which were used did not relate directly to interlayer forces. A high proportion of the work measured in a peel test is used in deforming the materials and the rate dependence is a consequence of the involvement of the rheological properties of the adhesive film in the gross effect measured.

Diffusion theory

The diffusion theory of adhesion has been advanced particularly by the Russian workers Voyutskii[13] and Vasenin[14] in more recent years and is essentially applied to the union of high polymers. It has the interesting facet that it emphasises the essential similarity of origin of the strength of a joint and the strength of a single material. The fundamental concept is that adhesion arises through the interdiffusion of the adherend and the adhesive. It has principally been applied to joints involving polymeric materials and it is difficult to see how it could be extended to include many other kinds of adherends.

It is based upon some of the most fundamental features of high polymers; the chain nature of the structure with the consequent flexibility and the ability of the chains to undergo Brownian movements on a sub-molecular scale.

Generally the adhesive molecules are more mobile and play a more active role in the diffusion process. However, when the adhesive is applied in solution, as it frequently is, and if the adherend is sensibly soluble in the solvent the substrate molecules will also diffuse to an appreciable extent into the adhesive layer. Overall, whether by an essentially one-way diffusion of adhesive molecules or by a two-way diffusion of both adhesive and adherend molecules, the clear-cut boundary between the two phases disappears and is replaced by a layer representing a gradual transition from one polymer to the other. The point of supreme and fundamental importance is that the mechanism of adhesion has become a three-dimensional, volume process rather than a two-dimensional, surface process.

This diffusion of the polymer molecules is clearly nothing more nor less than a mutual solubility which has been tacitly assumed by a number of workers who all stressed the correlation between compatibility of the polymers and the formation of good adhesive bonds.

A quantitative interpretation of this mechanism cannot be precisely given but some useful approximations can be made. If two compatible polymers are in contact then, as a result of thermal motion, interpenetration of segments of their macromolecules occurs. The segments which have diffused in this way will be held in their new positions by intermolecular forces. Thus the strength of the joint will depend upon the number of molecular chains which cross the phase boundary, the depth of their penetration and their individual strength, and these three factors may be considered separately.

$$F = f\,Q\,n$$

where F is strength of joint

f is strength of a single intramolecular bond

Q is number of molecular chains crossing boundary

n is depth of penetration measured in units of atomic grouping ($—CH_2—$)

Thus three separate problems arise:

(1) To determine the strength of a single bond, it is necessary to know the nature and position of the interacting groups and to have a considerable understanding of their quantum mechanical behaviour. Since our knowledge of these factors is generally fragmentary, this can only be estimated very approximately.

(2) The number of molecular chains crossing the phase boundary per square centimetre can be estimated by

$$Q = \left\{\frac{2N\rho}{M}\right\}^{\frac{2}{3}}$$

N = Avogadro's Number
ρ = density
M = molecular weight

provided that it is assumed that:

(i) the molecules have a linear structure and are arranged in a random manner

(ii) no special treatment is used to impose orientation upon the molecules

(iii) the diffusion occurs only with molecules in the surface of the polymer

(iv) only the end segments of the molecules are involved in movement, the centre parts remaining relatively stationary.

(3) Determination of the depth of penetration of the molecules is a complex problem which may be attempted on the basis of Fick's Law of Diffusion, and an expression may be obtained:

$$n = \frac{\{\pi\, K_D\, t^{(1-\beta)}\}^{\frac{1}{2}}}{2\, l \cos \frac{\alpha}{2}}$$

K_D is coefficient of diffusion of molecules.
β is rate of change of K_D with time.
l is bond length.
α is $(\pi - \phi)$ where ϕ is bond angle.
t is time.

These together will give

$$F = \frac{\pi^{\frac{1}{2}}\,(2N)^{\frac{2}{3}}}{2\, l \cos \frac{\alpha}{2}}\left\{\sum_1^2 f\left(\frac{\rho}{M}\right)^{\frac{2}{3}}\right\} .\, K_D^{\frac{1}{2}} .\, t^{(1-\beta)/2}$$

which to a first approximation can be written

$$F = C\, t^{\frac{1}{2}}$$

which corresponds, at least qualitatively, with the experimental data for autohesion and adhesion of rubbers.

Unfortunately most of the experimental data which might be

available for use in discussing these theories have been obtained by peel testing—a point to which I shall return. However, this leads to a Specific Work of Peeling (A) rather than the Strength of the joint (F). A further extensive theoretical treatment leads to an expression for this:

$$A = \frac{\pi^{\frac{1}{2}}(2N)^{\frac{2}{3}}}{4\epsilon \cos \frac{a}{2}} \cdot \left\{ \sum_{1}^{2} f_c^2 \left(\frac{\rho}{M} \right)^{\frac{2}{3}} \right\} K_D^{\frac{1}{2}} . \mathrm{t}^{(1-\beta)/2}$$

where ϵ is modulus of elasticity of a single chemical bond

f_C is force deforming one bond.

While there are difficulties in estimating the values for some of these, Vasenin[14] has obtained results for joints of polyisobutylene which appear to agree with this, as may be seen in Fig. 2.

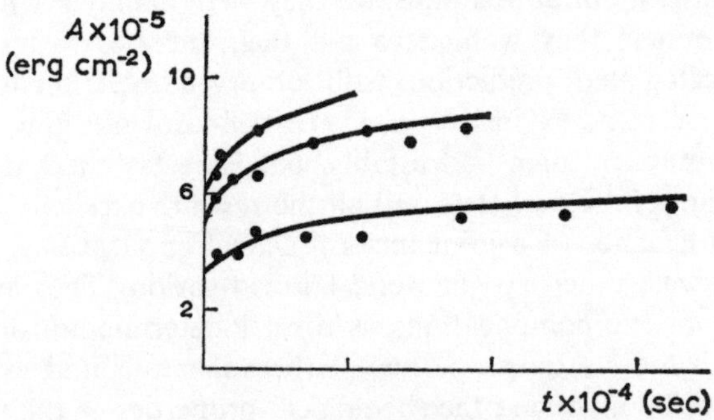

Fig. 2. Dependence of work peeling (A) upon time (t) for fractions of polyisobutylenes of different molecular weights. Curves are calculated, points are experimental. (From R. M. Vasenin's paper in *Adhesion—Fundamentals and Practice*, published for the Ministry of Technology, by Maclaren, London)

It is recognised that in order to support a diffusion theory of adhesion direct—as opposed to indirect—experimental proof is highly desirable, since all the indirect evidence may also be used to support alternative theories.

The criticism of this theory stems largely from the fact that all the evidence so far adduced for it is based on peeling tests with their essential ambiguity. But more important, it is very difficult to see how this theory can usefully contribute to an understanding of the adhesion of polymers to, for example, metal or glass.

Criticism and suggestion

So far this paper has tried to present and review other people's ideas, theories and criticisms as fairly as possible. Now I want to advance two particular comments of my own.

First of all I believe that there has been a good deal of confusion of purpose in most of the mechanical testing which has been carried out. It all began with simple, practical men who needed to fasten things together and so had to measure the strength of the joints which they were making. If instead of nailing or screwing two pieces of wood together, they were going to stick them, then a lap joint stressed in approximately the way it would be used was obviously the right and proper solution. However, before long scientists began to be interested in and concerned about adhesion. They began to 'dream dreams and see visions'[15] and then to advance theories about the how and why of adhesion. Since they were mainly scientists and not philosophers they wanted to put their theories to test and to correlate calculated prediction with observed experimental facts. They took over and extensively used the well-known testing methods which the practical men had established. Here lay the basic error; for while lap joints, peel tests and all the rest are excellent measures of the performance of a joint in its practical applications, they are not at all well suited to theoretical investigation. They introduce various geometric complications, add on indeterminate forces concerned with the elastic properties of the adherends and extend the complexities by involving the rheological properties of the adhesive.

The fundamental need is for obtaining data to correlate with theoretical predictions in tests specifically designed for that purpose and no other. Simple tensile tests of the separation of two parallel plates should lead to far more useful results than the more usual and practically useful tests so commonly applied or misapplied for theoretical calculation.

Now turning to the other idea. Chemists will be familiar with the technique used in discussing molecular structures. In order to describe a real structure, which in some respects is rather like but not identical to an ideal model structure A; and in other respects is similar but not identical to another ideal model structure B, an expression is used of the form $\Psi = M\psi_A + N\psi_B$ where ψ_A, ψ_B would describe the two models. M, N are mixing constants.

Thus in one expression both partial truths are expressed and combined and further no longer appear to be in opposition or mutual contradiction. Similarly, it seems to me that in discussions of adhesion there have been a number of different approaches each of which contains some particular insight into the total phenomenon. So far they have been considered to be in opposition or contradiction to each other and mutually exclusive. A more profitable and useful method might be to combine them and to represent the real situation in this way:

$$\Psi = \alpha\psi_{M} + \beta\psi_{A} + \gamma\psi_{D} + \delta\psi_{E} + \dots \omega\psi_{X}$$

where ψ_M is mechanical component of adhesion
ψ_A is adsorption " " "
ψ_D is diffusion " " "
ψ_E is electrical " " "
$\alpha, \beta, \gamma, \delta$ are mixing constants.

In most cases we might reasonably expect that α, β, δ will all have real and significant values but in some particular cases where the contribution of a component is negligibly small then the corresponding value of the constant would become zero. Thus we get away from the idea that various different theories are mutually exclusive and provide a framework into which all of them can reasonably be fitted.

Acknowledgement

I am most grateful to Dr. W. C. Wake of the Rubber and Plastics Research Association for his very considerable help in preparing this survey.

References

1. McBain, J. W. and Hopkins, D. G., *J. Phys. Chem.*, **29**, 88, 1925.
2. Maxwell, J. W., *Trans. Amer. Soc. Mech. Engs.*, **67**, 104, 1945.
3. *The Surveyor and Municipal and Country Engineering*, **121**, (3656), June, 1962.
4. Borroff, E. M. and Wake, W. C., *Trans. Inst. Rubber Industry*, **25**, 19, 1949.
5. Allen, K. W., *Aspects of Adhesion 1*, p. 11, University of London Press, 1965.
6. Ottewill, R. H., *Aspects of Adhesion 2*, p. 9, University of London Press, 1966.
7. Eley, D. D., *Adhesion*, Oxford University Press, 1961.
8. Schonhorn, H., *Adhesion: Fundamentals and Practice*, Ministry of Technology, Maclaren, London, 1969.
9. Bright, K. and Malpass, B. W., Paper Presented to British Polymer Physics Group Conference, 6–7 June, 1967, Private Communication. *European Journal of Polymer Science*, 1968.

10. Deryaguin, V., *Research*, **8,** 70, 365, 1955.
11. Jeans, J. H., *The Mathematical Theory of Electricity and Magnetism*, p. 83, Cambridge University Press, 1925.
12. Skinner, S. M., Savage, R. L. and Rutzler, J. E., *J. Appl. Physics*, **24,** 438, 1953.
13. Voyutskii, S. S., *Autohesion and Adhesion of High Polymers*, p. 138, Inter-science, 1963.
14. Vasenin, R. M., *Adhesion: Fundamentals and Practice*, Ministry of Technology, Maclaren, London, 1969.
15. Joel ii. 28.

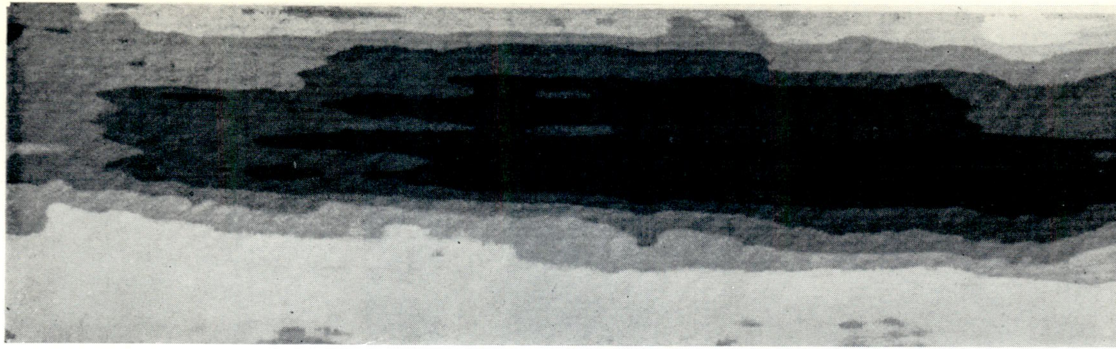

PLATE 3 Contoured recording obtained with the Harwell ultrasonic interference micrometer

A Review of Bond Testing Techniques

R. S. SHARPE, B.Sc., F.Inst.P., F.I.M.

NONDESTRUCTIVE TESTING CENTRE, A.E.R.E., HARWELL

Modern materials and fabrication techniques are continually coming up against problems of interface integrity and presenting inspection requirements of varying complexity. For the purpose of this review I am ignoring welded interfaces as this is a rather specialised form of bonding which, I imagine, is well beyond the terms of reference of this meeting, although even here there are many inspection techniques and philosophies which are of general interest to other forms of bonding.

Apart from welding, surfaces can be joined directly, with a metal interlayer such as braze or solder, or with an adhesive. The interfaces can constitute a simple lap joint or be part of a composite structure such as a honeycomb construction. The interface may be between similar materials, such as two metals, or between widely differing materials such as rubber and metal, propellant and metal or cermet and metal. The interface may be a macro constructional joint or it may, on a micro scale, be an integral property of a composite material such as a fibre or whisker reinforced matrix, where the matrix itself may be a metal, a plastic or a ceramic (Fig. 3).

The direct metal to metal bond and its inspection problems have received considerable attention in the nuclear fuel element industry; adhesive bond technology has largely come into prominence through the requirements of the aircraft industry; the bonding together of different materials has, to quote but one example, certainly occupied the attentions of solid rocket manufacturers in a big way, and the characteristics of microbonds in composites is very much a topical problem.

In all of these cases the common feature from an inspection point of view is assessing in some way the integrity and efficiency of an interface joint. With the limited number of techniques and physical principles that one can use it is, of course, not surprising that parallel approaches have developed, and from the scientific point of view it

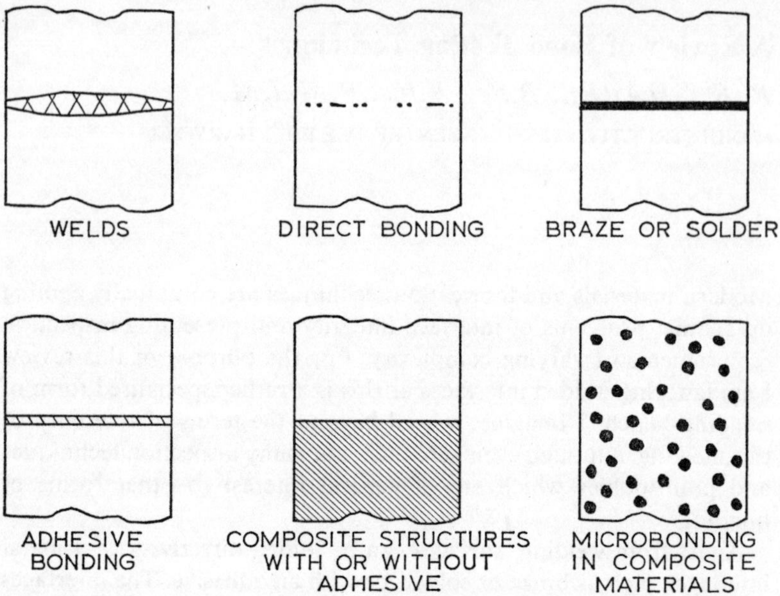

Fig. 3. Types of interface for which bond testing techniques have been developed

is unfortunate that the sideways lines of communications have sometimes been a little clogged, by problems of commercial and defence security.

The requirements demanded from a bonded interface vary with the application and this fact has naturally modified the inspection approach. Strength of bond is, to a greater or lesser extent, generally the underlying requirement. In the case of nuclear fuel elements, however, strength goes hand in hand with thermal efficiency as joint criteria for a satisfactory bond. In the case of solid fuel rockets the completeness of the bond is in itself a very relevant factor in preventing uneven burning. This simpler concept of whether or not a bond exists is also of interest in devices such as ultrasonic transducers where variations in the back damping of the transducer plate caused by variations in bond efficiency can affect the output distribution of energy (Plate 1). Again, in some cases the shape of the unbonded area is a significant feature and this may become especially important with microbonds in extruded composites.

Measuring the strength of a bonded joint is a notorious 'bête noir' of the inspection world. No completely satisfactory solution has yet been found. This is perhaps not surprising since no satisfactory solution has yet been found to the problem of nondestructively monitoring the mechanical strength of a *solid* piece of metal.

One is consequently obliged to approach the problem indirectly to try to find a property which is directly related to strength and which does lend itself to nondestructive monitoring.

The most obvious secondary characteristic to work on is the extent of the bond. It is reasonable to assume that if parts of the interface are not physically joined together the strength will be proportionately reduced.

Ultrasonic methods came into their own in detecting this type of physical 'unbond' since an internal interface with air on one side has such a high acoustic impedance that it is virtually opaque to ultrasound. Transmission or reflection methods have been used to detect such a barrier and a large amount of ingenuity has been exercised, especially in the nuclear and rocket fields to develop sensitive techniques and sophisticated methods of recording, analysing and displaying the results. Unfortunately, as with most tests, there is no simple distinction between 'black' and 'white'. Resolution and sensitivity inevitably depend on the test conditions and factors other than degree of 'unbond' may significantly influence the results (Fig. 4).

Perhaps the most disturbing feature of this particular test, however, is the fact that if the two surfaces are in intimate contact—but definitely unbonded in the accepted sense—ultrasonic energy can be propagated across the interface as though it were a bond with good mechanical strength (Plate 2). Experimental results suggest, moreover, that with some surfaces a separation of up to about one micron presents no effective barrier to ultrasound. Admittedly the technique is extremely effective in detecting seriously defective bonds, but the limitations should be very clearly realised so that the results can be interpreted with adequate discretion.

Variants of the ultrasonic propagation technique include the excitation of Lamb waves and compressional resonances, both of which can only be supported if the inner surface is unbonded and can vibrate independently. At some interfaces a phase change can also be used to detect unbonded areas.

The ultrasonic camera was specifically developed in a number of centres with the particular application to bond testing in mind. It

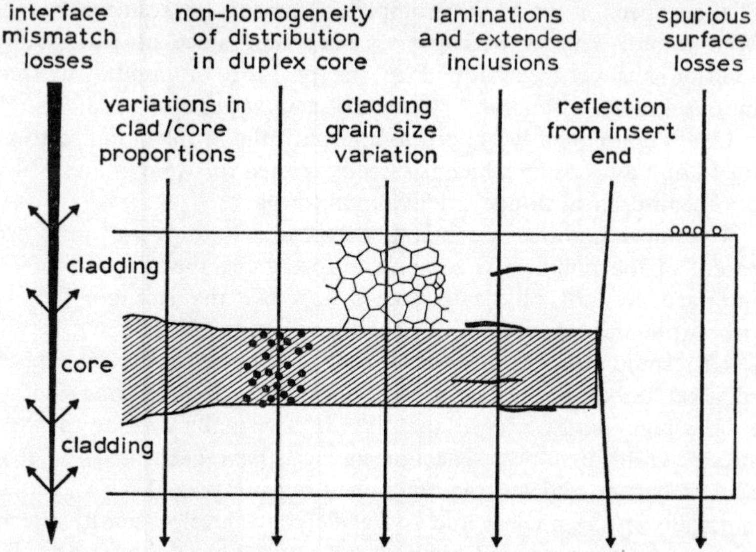

Fig. 4. Factors that can influence bond testing in nuclear fuel plates

provides a higher inspection speed than is possible with other ultrasonic techniques and has the added advantage of providing a direct visual display of any unbonded areas. However, it still suffers from the inherent limitation already referred to which is common to all ultrasonic techniques. In the rocket field, the specification generally calls for a less sensitive technique than the nuclear requirements demand but other problems associated with the design and with the acoustic properties of the propellant have led to air coupling techniques and to more sophisticated forms of data processing and display.

The widespread introduction of adhesive bonding led to parallel developments in a number of the above areas of inspection. A new possibility however, presented itself, since the vibrational characteristics of two metal plates joined by a viscoelastic adhesive layer of finite thickness are influenced by the elastic constants of the adhesive and the mass and stiffness of the vibrating system. These, in their turn, are related to the cohesive strength of the adhesive and hence to the overall mechanical strength of the bonded joint. There is a vast literature on this subject, containing a wealth of documented experience with commercially available instruments designed to operate on this principle.

In operation a probe is placed on the surface of the bonded sheet and vibrated by means of a swept frequency oscillator. The resonance frequency of the vibrating system (probe plus bonded sheet) and the amplitude of the resonance are displayed. Variations in these two parameters are used as a comparative measure of bond 'quality'. It is essential to calibrate the instrument against a series of control samples of known bond strength prepared under identical conditions of surface treatment, adhesive thickness, curing time and curing temperature. The techniques are undoubtedly of immense practical importance (and the protagonists are equally enthusiastic!), but they suffer from the fact that they are based on a mass of empirical data and practical experience rather than on a complete scientific understanding of the physical and mechanical inter-relations of the bonds, and of the acoustic principles involved in analysing composite vibrators. This is undoubtedly an area of bond testing that would benefit from a controlled programme of scientific investigation.

As with other ultrasonic tests there are limitations. Whilst variations in the physical characteristics and porosity of the adhesive layer are easily detected by their influence on the resonance, variations in

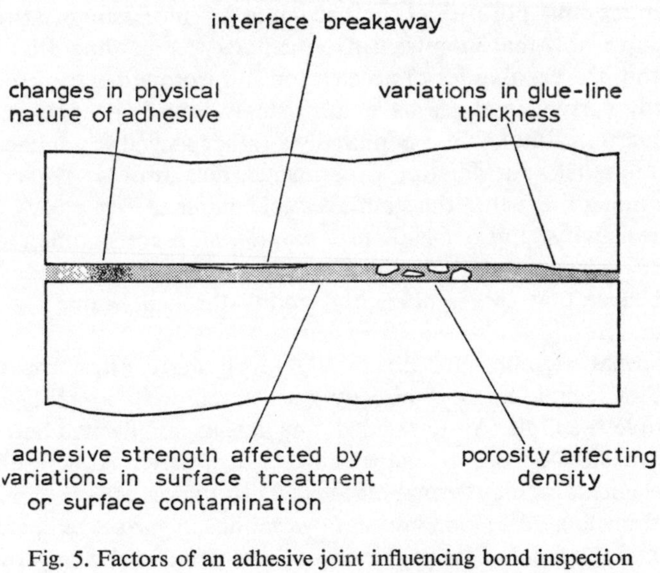

Fig. 5. Factors of an adhesive joint influencing bond inspection

adhesive strength at the interfaces are not detectable (Fig. 5). An improved form of ultrasonic resonance technique has been developed at Harwell which relies on exciting a metal sheet into longitudinal resonance. It can measure the resonance to an accuracy of at least 0·1 per cent and present the value as a digital output. Initial tests on lap bonded sheets have shown that the resonances of both sheets can be excited. Although the effect of bond variables on these resonances has not yet been checked, this might provide the basis for an additional form of ultrasonic test with advantages possibly of higher sensitivity and higher speed of inspection. Being a non-contact method it has the advantage of being able to present the resonance frequency as a contoured recording (Plate 3). This is a technique that should be examined further in this context.

If ultrasonic techniques are not the complete answer to bond testing what other possibilities exist?

Radiography has little to offer since laminar defects are orientated in the direction of minimum sensitivity. Brazed bonds, however, have been inspected in this way (especially with the honeycomb type of construction), to detect variation in braze distribution. In fact, direct view X-ray techniques have been specifically developed with this application in mind. In the case of adhesive bonds, neutron radiography might prove an effective method of locating density variations and porosity. It is becoming an increasingly attractive technique now that more sensitive methods of recording are putting it within the capabilities of accelerator and isotope neutron sources.

Eddy current methods have surprisingly found an application in bond testing although it is admittedly a rather specialised application. In some nuclear irradiation experiments, heat transfer is increased by bonding the fuel to the stainless steel container with a NaK layer. The resistivity values of NaK and stainless steel are so different that an eddy current scanning coil can detect the small local voids in the bond layer that can significantly modify the heat transfer characteristics.

Thermal methods are potentially much more attractive and a range of techniques have already been developed and applied. The principle is simple. An unbonded area acts as an effective barrier to heat transfer and the techniques have been widely used with nuclear fuel elements where thermal efficiency is so important. Variations in heat transfer across a bond result in variations in surface temperature. The main technological problems are associated with the provision of a uniform and rapid heat source and in devising sensitive methods of monitoring very small temperature differences.

Methods based on directly measuring the temperature have been devised. De Bruyne used a heat source flanked by two thermocouples. The NPL and the AID thermal comparator with a differential thermocouple system to measure effective thermal conductivity of a surface are more recent development that can be used to make similar measurements. Other techniques have made use of surface coatings with sharp melting points, thermographic phosphors and cholesterol coatings. Direct infrared photography can detect temperature differences when the surface temperature is high (around 300°C). Infrared detectors using scanning optics are considerably more sensitive and have been reported as being capable of detecting temperature differences as small as 0·001°C on surfaces only slightly above room temperature. More recently infrared sensitive Vidicons have been developed for direct presentation of thermographs on a television monitor screen but the contrast sensitivity so far achieved is only about 3°C. Although the thermal methods so far described are limited to detecting voids and unbonded interfaces heat stressing can be used very effectively to monitor bond strength. I do not intend to dwell on this area as much of the work in the U.K. has been carried out by Thompson and his colleagues at A.I.D. and will, I am sure, be discussed in their paper later in the Conference. Our own experience with fuel plates is the so-called 'blister' test. A poorly or weakly bonded joint can raise a permanent blister if there is sufficient air or gas entrapped locally to expand and plastically deform the clad layer. The A.I.D. type of technique is more sophisticated in that surface displacements far smaller than those required to raise a blister are detected.

The thermal stressing techniques provide in some ways a far more positive measure of bond strength than the ultrasonic or heat transfer techniques. I suspect however, that the ability to detect small areas of defective bond is not too high and they can be criticised in that the test itself may weaken a bond to the detriment of its subsequent performance.

There is, of course, a wide range of possible techniques to measure the very small surface displacements produced when a weak bond is stressed. Strain gauges, air gauges and capacitance gauges have been used. Photoelastic coatings might be effective in some instances and laser holography can reveal interference fringes from displacement contours each separated by only a wavelength of light, but I have not seen the method applied particularly to bond testing.

Other methods of stressing have been developed. The vacuum cup technique can be used to detect grossly defective bonding and high energy sonic and ultrasonic energy has also been used to disrupt and separate weak and defective interfaces. In the former case a simple displacement was used but in the latter the high energy ultrasonic energy was modulated by a lower energy but higher frequency ultrasonic probing beam as in conventional pulse echo testing.

One of the earliest methods of bond testing—and still an effective one—is to tap the bond with a cane or pencil and listen for variations in the 'ring'. This, surprisingly, is a technique that has never been really scientifically developed and yet is one where there is a wealth of potential application, not only in this context but in many other areas of nondestructive testing.

An interesting variant of this acoustic method is to allow the bond to generate its own sonic or ultrasonic emission under stress and to interpret this sound pattern in terms of bond characteristics. Pollock has observed this effect with an adhesive bond incidentally in a basic study of sound emission. It is undoubtedly an observation that calls for more detailed study. More extensive experiments at the G.A.T. Corporation in America have shown that emission above 16 kHz increases considerably as the applied stress approaches the ultimate strength of a bond. Other work has also led to a realisation of the potential of this technique in detecting the failure of micro-bonds in fibre reinforced composites.

In any sonic emission test, the ambient noise must be reduced to a minimum. Pollock has designed a hydrostatic method to load his bonded samples but a more flexible approach might be to use a thermal stressing technique and this would seem to be a very promising line for further development.

Comments made by Mr. A. Pollock at the end of the lecture

Acoustic emission is a strength-related side effect occurring during deformation. Bond testing methods based on the effect are being developed. Acoustic emissions are sounds generated in materials as deformation takes place. When deformation is accompanied by the local release of energy (as for example during crack propagation) a fraction of the energy is carried away in a stress wave which leaves the source of damage and may be detected as an acoustic emission.

It is evident that acoustic emission tests are damaging in that the method relies on deformation occurring during the test, and cannot be used to detect static defects (such as non-bonded areas) already present in the specimen. The method may be classed as non-destructive insofar as the integrity of the test body may not be immediately threatened although some deformation has taken place. The method is particularly relevant to large test bodies where a measure of plastic deformation is an integral part of the specimen's accommodation to test or service conditions.

The acoustic emission test is closely related to the conditions of service. Stress is applied mechanically or thermally and the response of the specimen to the stress is observed. The advantages of non-destruction are thus combined with the directness of the destructive test.

The equipment needed to observe acoustic emissions is not complicated. It consists simply of transducer, amplifier and recording or counting system. In large structures several transducers can be used, mounted appropriately near the various bonded areas. Acoustic emissions are often of very low amplitude and their discrimination from background noise needs care in industrial applications; in the laboratory this does not present much of a problem. Filtering helps greatly since emissions are generally of much higher frequency (tens of kilocycles) than background noise. The frequencies observed in the emissions are governed by the vibrational characteristics of the specimen and transducer.

Many mechanisms give rise to emissions. In brittle resins, cracking is a major source. Frictional noises may be important and tearing or peeling in flexible (e.g. rubber-based) adhesives is also a 'noisy' process.

Two methods of data analysis can be used to obtain information about the performance of the bond under test. Both these methods must at present be empirically based.

First, the level of acoustic activity can be monitored. Typically, emissions would be counted. As a test proceeds, emission activity follows general trends which can be related to bond performance. For instance, emission activity shows a characteristic increase prior to failure. The relationship of early acoustic activity to ultimate bond strength has not, to the author's knowledge, been studied. On a more sophisticated level, emission activity can show whether micro-deformation events occur independently or co-operatively.

The second method of data analysis is related to the frequency of the emissions. This method is related to the Fokker bond testing method and thus assesses the cohesive but not the adhesive strength of the bond. Specimen resonances are excited not by externally mounted transducers (as in the Fokker method) but by the acoustic emissions themselves. The change in emission frequency as a test proceeds gives a measure of the decreasing cohesive strength of the bond as it undergoes damage. This method has been found useful in measuring adhesive bond strengths on metal honeycomb core panels.

There is evidently plenty of scope for the development of bond testing methods based on the acoustic emission phenomenon.

Conclusion

Adhesive bond testing presents problems very similar to those experienced with other types of bond and it would seem that many bond testing techniques are being developed in parallel fields without too much cross fertilisation of ideas. Undoubtedly much has been achieved by empirical approaches, but equally it is easy to see that much more could probably be achieved by a scientific study of the techniques already being applied and by the controlled development of some of the newer approaches that have been suggested.

There is never likely to be a single technique to cater for all the requirements of bond testing and we are some way from achieving the optimum test for monitoring bond strength. However, with the amount of effort being devoted to the problem, we shall soon arrive at the inevitable impasse of the testers overtaking the evaluaters so that the specifications are set by the sensitivity of the testing techniques rather than the demands of service performance.

Summary

The paper examines bond testing in the broadest sense and shows the range of nondestructive techniques which have been developed for different types of bond and for different types of bond application. The problems of adhesive bonds have led to an upsurge of experimental development and empirical relationships have evolved which suggest that, with reservations, the strength of the bond can be monitored. It is interesting to contrast the optimism of those working in this field with the more resigned approach of those concerned with the similar problem of testing the strength of metal/metal interfaces in other fields. In bringing together the testing techniques and testing philosophies of both fraternities, exchange of experience may suggest new lines of development and new areas of development which are of common interest. The paper includes a selective review of the literature and comes to the inevitable conclusion that bond testing is one of the many areas of nondestructive testing that urgently requires scientific laboratory investigation to supplement and co-ordinate the mass of experimental data and operational experience now available.

References

A list of references classified under techniques has been prepared by the N.D.T. Centre Information Service to support this review. No specific correlation is made with the text of the review.

Bibliographies

1. Non-destructive testing of adhesive bonded joints. North American Aviation Inc. Rep. SID 62–26 (=AD–282, 461) 1962. (33 refs.) —a partially annotated bibliography.
2. Meckelburg, H. and Althof, W., Methods of non-destructive testing of adhesive bonded metal joints. DFL–BER (216) 1963. (32 refs.) Translation TIL/T 5679 by Min. of Aviation, Sept., 1966.
3. Markin, Y. I. and Voyutskii, S. S., Methods for testing the strength of adhesive joints. *Ind. Lab.*, **29,** 867, 1963. (49 refs.) —mostly destructive testing.
4. Non-destructive testing theory and application for composites. Bibliography. North American Aviation Inc., Oct., 1965. 219 pages. (Contracts NAS 8–11733, AF 33 (615) -2865.)

General

5. Blackwell, R. F., Inspection and testing of honeycomb and honeycomb sandwich bonded aircraft structures. *Proc. Bonded Aircraft Structures Conf., Cambridge*, 1957. CIBA (ARL) Ltd.
6. Jeffery, F. D. C., The inspection of bonded aircraft structures. *Proc. Bonded Aircraft Structures Conf., Cambridge*, 1957. CIBA (ARL) Ltd.
7. Filippi, F. J., Qualitative control of brazed sandwiches. *Nondestr. Test.*, **17,** 39, Jan.–Feb., 1959.
8. McMaster, R. C., Summary of the SNT Symposium on steel brazed honeycomb structures. *Nondestr. Test.*, **17,** 263, Sept.–Oct., 1959.
9. Schlegel, H., Non-destructive testing of adhesive bonded metal joints. *Fertigungstechnik Betrieb*, **10,** (12), 751, 1960.
10. Aproian, P. and Woollett, R., Dynamic strength testing of adhesives and piezoceramics. Abstract of paper at 59th Meeting Acoustical Soc. America. *J. Acoustical Soc. America*, **32,** 939, 1960. —stresses with half wave resonator and strains measured.
11. Cusick, J. M., Non-destructive testing of large solid propellant case-bonded rocket motors. *Nondestr. Test.*, **18,** 195, May–June, 1960.
12. Bonding Inspection: Book issued by Federal Aviation Agency, July, 1960.
13. Gonzalez, H. M. and Cagle, C. V., Non-destructive testing of adhesive-bonded joints. *ASTM STP* 360, 148, 1963.
14. Martin, G. et al., Development of non-destructive testing techniques for honeycomb heat shields. North American Aviation Report NA–65–452, June, 1965.
15. Moore, J., Development of non-destructive testing techniques for Saturn honeycomb heat shields. North American Aviation Report NA–65–618, Aug., 1965.
16. Schmitz, G. and Frank, L., Non-destructive testing for evaluation of strength of bonded material. NASA–CR–67983 (Contract NAS8–11456), Sept., 1965. —includes acoustic emission work.

Ultrasonic resonance techniques

17. Dietz, A. G. H., Closmann, P. J., Kavanagh, G. M. and Rossen, J. N., The measurement of dynamic modulus in adhesive joints at ultrasonic frequencies. *ASTM Proc.*, **50,** 1414, 1950 and *ASTM STP* 101, p. 117, 1949.
18. Noble, A. W. and Westervelt, P. J., A resonant bar method for determining the elastic properties of thin lamina. *J. Appl. Phys.*, **21,** (4), 304, 1950.
19. Dietz, A. G. H., Bockstruck, H. N. and Epstein, G., Non-destructive determination of mechanical properties and deterioration of adhesives. *ASTM STP* 138, 1952.
20. Jayne, B. A., A non-destructive test of glue bond quality. *Forest Prod. J.*, **5,** (5), 294, 1955.
21. Grindrod, J., Testing adhesive bonded metal–sandwich structures. *Sheet Metal Ind.*, **32,** 338, 1955.
—describes use of Stubmeter.
22. Grey, E. L., Adhesive bonded joints tested by ultrasonics. *American Machinist*, **100,** (27), 122, 1956.
—describes use of Stubmeter.
23. Arnold, J. S., An ultrasonic technique for non-destructive evaluation of metal-to-metal adhesive bonds. *ASTM STP* 213, 79, 1956.
—describes work on Stubmeter.
24. Tapp. P. F. et al., Measurement of adhesive moduli. *Aircraft Eng.*, **29,** (345), 350, 1957.
25. Schijve, J., Ultrasonic resonance testing of glued metal joints. *Aircraft Eng.*, **30,** (355), 269, 1958.
—describes use of Vidigage and gives theory.
26. Vincent, C. T. and Arnold, J. S., Ultrasonic evaluation of structural adhesive bonds. *Proc. Nat. Electronics Conf.*, **14,** 951, 1958.
—describes work with Stubmeter.
27. Arnold, J. S., Development of non-destructive tests for structural adhesive bonds. WADC Tech. Rept. 54–231, Pt. 1 Oct., 1954 to Pt. 7 Sept., 1959.
28. Adhesive bond testing. *Aircraft Prod.*, **22,** (2), 75, 1960.
—production application of Fokker bond tester.
29. Kanamaru, K. et al., Non-destructive quality test of adhesive joints by means of resonance measurement. *Kolloid Z.*, **170,** (2), 123, 1960.
30. Schliekelmann, R. J., Quality control in metal bonding. *Industrie-Anzeiger* **83,** (34), 525, 1961.
—describes use of Fokker bond tester.
31. De Jonge, J. B., The behaviour of the Fokker bond tester on double lap joints. NLR–TN M 2098, March, 1962.
32. Hartmann, A. and de Jonge, J. B., Non-destructive and destructive tests on Redux bonded single and double lap joints with various glue-line thicknesses. NLR–TN M 2099, March, 1962.
33. Boruff, Von H., Evaluation of ultrasonic devices for inspection of adhesive bonds. Martin Report ER 10911–9, Aug., 1961 and Quart. Prog. Report. No. 10 Martin Marietta Co., Jan., 1962.
—comparison of commercial equipments.
34. Miller, N. B. and Boruff, Von H., Adhesive bonds tested ultrasonically. *Adhesive Age*, **6,** (6), 32, 1963.
—comparison of commercial equipments.
35. Gonzalez, H. M. and Cagle, C. V., Non-destructive testing of adhesive-

bonded joints. *ASTM STP* 360, 148, 1963.
—describes results with Fokker bond tester.

36. Schliekelmann, R. J., Non-destructive testing of adhesive-bonded metal structures. *Adhesives Age*, **7**, (5), 30, 1964 and **7**, (6), 33, 1964.
37. Hertz, J., Investigation of bond deterioration by use of Fokker bond tester. GD/A–ERR–AN–682, Dec., 1964.
38. Botsco, R., Acoustical testing of honeycomb structures from one accessible side (Resonance Technique). North American Aviation Report CD–3804, June, 1964.
39. Evans, G. B., The Fokker Bond Tester. *Sheet Met. Ind.*, **42**, (462), 751, 1965.
40. Smith, D. F. and Cagle, C. V., Ultrasonic testing of adhesive bonds using the Fokker bond tester. *Mat. Eval.*, **24**, 362, July, 1966.

Other acoustic and ultrasonic techniques

41. Baumeister, G. B., Production testing of bonded materials with ultrasonics. *ASTM Bull.*, (204), 50, Feb., 1955.
42. Schijve, J., Investigation on the ultrasonic testing of glued metal joints. National Aero Res. Inst. Report M1995, Amsterdam 1955.
43. Bernert, J. and Bonisch, S., Non-destructive testing of glued joints by ultrasonic methods. *Fertigungstechnik Betrieb*, **7**, (2), 81, 1957.
44. Worlton, D. C., Ultrasonic testing with Lamb waves. *Nondestr. Test.*, **15**, 219, July–Aug., 1957.
45. Kammerer, C. C., Ultrasonic inspection of adhesive-bonded aluminium sandwich structure. *ASTM STP* 278, 47, Oct., 1959.
46. Hughes, E. T. and Burstein, E. B., The evaluation of bond quality in honeycomb panels using ultrasonic surface wave techniques. *Nondestr. Test.*, **17**, 373, Nov.–Dec., 1959.
47. Erdman, D., Ultrasonic pulse-echo techniques for evaluating thickness, bonding and corrosion. *Nondestr. Test.*, **18**, 408, Nov.–Dec., 1960.

47a. Sharpe, R. S., Aveyard, S. and Sayers, J. F., The inspection of fuel element bonding. AERE–R 4605, 1964.

48. Kleint, R., McKown, R. and Ramsey, J., Application of ultrasonics to solid rocket systems. WADD TR 60–553, Oct., 1960.
49. Reese, J. P. and Boruff, Von H., Evaluation of ultrasonic test devices for inspection of adhesive bonds. Martin Co. Report ER 109011–1 Oct., 1959 to ER 10911–9 Aug., 1961.
50. Althof, W., Non-destructive testing of metal-adhesive bonds with ultrasonic pulse-echo instruments. DFL–BER (217) 1963. Translation TIL/T 5680 Min. of Aviation, Aug., 1966.
51. An ultrasonic technique for the evaluation of bonded honeycomb structural materials. MTP–P and UE–M–63–12.
—transmission technique.
52. Jacobs, J. E., Collis, W. J. and Berger, H., An evaluation of an ultrasonic inspection system employing television techniques. *Mat. Eval.*, **22**, 209, May, 1964.
53. Peters, B. F., An ultrasonic method for evaluating bonds in large metallic bearings. *Mat. Eval.*, **22**, 257, June, 1964.
54. Nanda, M. M., Sinclair, N. A. and Lovelace, J. F., Ultrasonic testing of brazed pipe joints. *Mat. Eval.*, **22**, 315, July, 1964.
55. Harris-Maddox, B., The ultrasonic testing of boost motors. *Ultrasonics*, **2**, 60, April–June, 1964.
56. Thompson, W. and Spear, J. A. N., An ultrasonic method of detecting

areas of non-adhesion in aluminium alloy honeycomb panels. AID Report NDT 3634C/64, Nov., 1964 (with acknowledgements to Ministry of Aviation).

57. Stinebring, R. C. and Harrison, R. H., Non-destructive testing of rocket components using microwaves and low frequency ultrasonics. *Mat. Eval.*, **23**, 17, Jan., 1965.
58. Clotfelter, W. N., Acoustic techniques for the non-destructive evaluation of adhesively bonded composite materials. NADA–TM–X–53219 (=N–65–25346) March, 1965.
59. Cook, K. V. and McClung, R. W., An ultrasonic technique for evaluation of heat exchanger brazed joints. *Mat. Eval.*, **24**, 97, Feb., 1966.
60. Harris-Maddox, B., The ultrasonic testing of solid-propellant rocket motors. *Mat. Eval.*, **24**, 101, Feb., 1966.
61. Botsco, R. J., Non-destructive testing of composite structures with the sonic resonator. *Mat. Eval.*, **24**, 617, Nov., 1966.
62. Lopilato, S. A. and Carter, S. W., Unbond detection using ultrasonic phase analysis. *Mat. Eval.*, **24**, 690, Dec., 1966.
63. Kammerer, C. C., Ultrasonic quality inspection of bonded honeycomb assemblies is automated. NASA Tech. Brief 66–10544. Nov., 1966.
64. Pollock, A., Acoustic emission. Paper at Harwell course on Advances in Non-destructive Testing. Feb., 1967. To be published.

Thermal techniques

65. Yetitto, P. and Gorman, R., Feasibility study of a non-destructive testing infrared inspection system for bonding flow detection. Perkin Elmer Corp. PE–TR–5959, April, 1961.
66. Thompson, W. and Josling, A. T., A non-destructive test for adhesive bonded joints. AID Report, NDT/1712, Jan., 1964 (with acknowledgements to Ministry of Aviation).
67. Green, D. R., Emissivity independent infrared thermal testing method. *Mat. Eval.*, **23**, 79, Feb., 1965.
68. Alzofon, F. E., An infrared non-destructive testing system for rocket motors. *Mat. Eval.*, **23**, 537, Nov., 1965.
69. Thompson, W. and Senior, H. A., A non-destructive test for adhesive bonded fibreglass–aluminium alloy joints. AID Report, NDT/56/65, March, 1965 (with acknowledgement to Ministry of Aviation).
70. Cade, C. M. and Hanwell, A. E., Thermal patterns. *Science J.*, **2**, (2), 62, 1966.
71. Berger, H. and Kraska, I. R., Characteristics of an infrared vidicon television system. *Mat. Eval.*, **24**, 197, April, 1966.

Eddy Current Techniques

72. Vandervort, P. S., Morris, J. W. and Fitch, S. H., Evaluation of eddy current inspection for hot sodium bonding in stainless steel clad uranium carbide reactor fuel elements. *Mat. Eval.*, **23**, 196, April, 1965.
73. Renken, C. J., A pulsed eddy current test system using reflected fields. *Mat. Eval.*, **23**, 622, Dec., 1965.

Radiography

74. Maszy, S., Radiography tests brazed stainless sandwich. *American Machinist*, Dec., 1956.

Direct strain measuring techniques

75. Rutzler, J. E., Effects of stress on mechanical and electrical properties of metals bonded by means of adhesives. *Case Inst. of Technology*, 1954.
 —uses resistance gauges to measure strains.
76. Norris, C. B., James, W. L. and Drow, J. T., A strain gauge for the measurement of strains in adhesive bonds. *ASTM Bull.*, (218), 40, Dec., 1956.
 —uses a capacitance measurement across the bond.
77. Redner, S., Non-destructive testing using the photoelastic coating technique. *Mat. Eval.*, **22,** 517, Nov., 1964.
78. Non-destructive test predicts strength of adhesive bonds. *Machine Design*, **37**, (5), March, 1965.
 —relates slope of stress/strain curve to bond strength.

Miscellaneous

79. Wall, M. R. Some suggestions for the use of holography in metrology. AWRE Pamphlet No. 1, Jan., 1967.

The Strength of a Lap Joint

L. GREENWOOD, B.Sc.
E.R.D.E., WALTHAM ABBEY

Introductory

The commonest type of joint in use today is the lap joint where two adherends overlap and are joined by an adhesive. It is the joint that is used in service and also for testing and will probably be so used for a very long time to come. The main advantage of this joint is that it is the natural one to make and involves least preparation from the design side. Further, where the surfaces need chemical preparation, the adherends are easy to handle. It is also a simple type of joint to cure.

The chief disadvantage of this type of joint is that the stress distribution is by no means uniform along the joint and very high peak stresses are produced at or very near the ends of the overlap. If the applied load is divided by the area of overlap this stress is the mean stress and the ratio of these peak stresses to this mean stress is known as the stress concentration factor and this is studied here.

There is evidence to support the very widely held theory that there is no such thing as an adhesive failure. This means that a joint will fail when the stress at any point in the glue line reaches the failing stress of the glue and this will clearly be at the point of maximum stress concentration. Further, any modification that we can apply to the design of joint which will reduce this stress concentration factor will increase the joint strength.

It is therefore probable that a careful study of the distribution of stress in a lap joint will indicate ways in which the strength of the joint can be modified.

Historical

The earliest theoretical analysis of lap joints is that done by

Volkersen.[1] The next analysis was that of Goland and Reissner.[2] Subsequent analyses have been carried out by Plantema,[3] Cornell,[4] Lubkin and Reissner[5] and Lubkin,[6] Mylonas and de Bruyne[7] and McLaren and McInnes.[8]

Lap joints general

When a lap joint like that in Fig. 6 is loaded, it is reasonable to assume that the uniform tension in A falls off linearly to zero at the free end

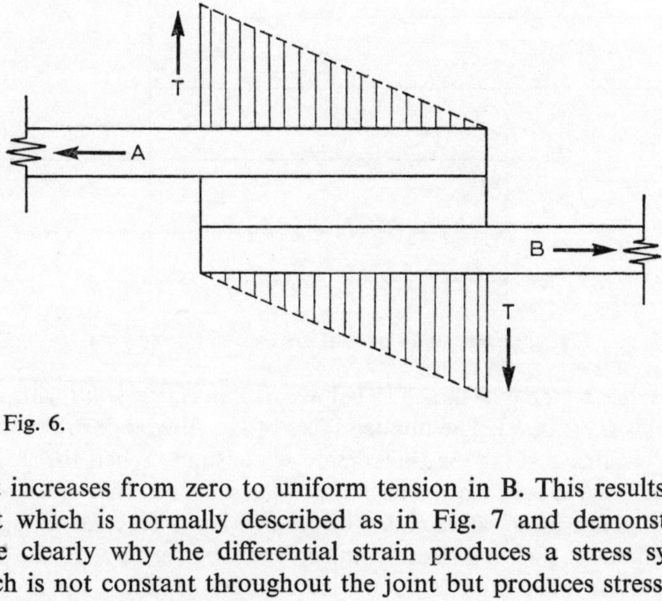

Fig. 6.

as it increases from zero to uniform tension in B. This results in a joint which is normally described as in Fig. 7 and demonstrates quite clearly why the differential strain produces a stress system which is not constant throughout the joint but produces stress concentrations at the two ends.

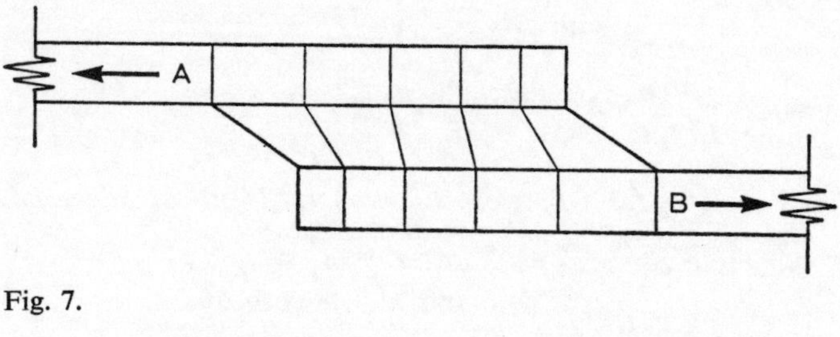

Fig. 7.

Stress distribution due to stretching

When Volkersen wrote his paper, he considered the stress distribution that develops in a lap joint under load due to stretching of the adherends only; he ignored the tearing stresses at the free ends. He considered two elastic adherends of uniform thickness t_1 and t_2 and Young's moduli E_1 and E_2 bonded over a length $2C$ by an adhesive of

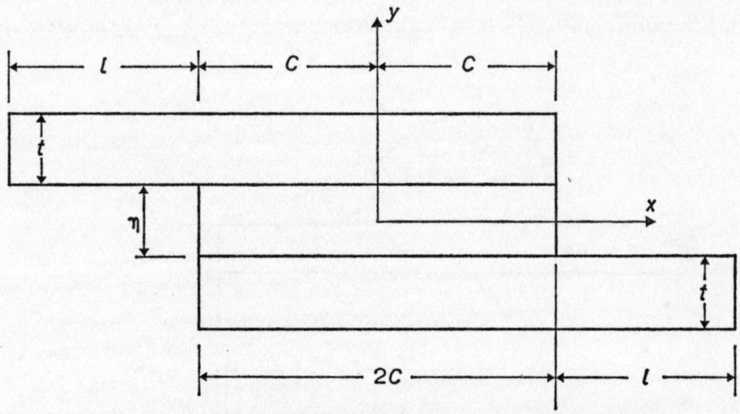

Fig. 8. Definition of parameters used in the lap joint

thickness η which is assumed to behave like an elastic solid with shear modulus G_a (Fig. 8). The bonded faces of the adherends are assumed to be parallel so that the thickness η is constant. When the bonded members are in pure tension as in Fig. 7 the shear stress in the adhesive layer is a maximum at each end of the overlap. Volkersen compared the maximum shearing stress at the end of the overlap with the mean stress and evaluated n, the stress concentration factor which is the ratio between these stresses and is

$$n = \frac{\delta}{\epsilon}\left\{\frac{2\epsilon^2 - 1 + \cosh 2\epsilon\delta}{\sinh 2\epsilon\delta}\right\}$$

where $\delta^2 = \dfrac{2C^2 G_a}{E_2 t_2 \eta}$; $\epsilon^2 = \dfrac{E_1 t_1 + E_2 t_2}{2E_1 t_1}$

(a) *The Case of a Joint Between Elastomeric Identical Adherends*

Here $E_1 = E_2 = E$ and $t_1 = t_2 = t$, i.e. $\epsilon = 1$
and $n = \delta \coth \delta$

If $E \sim 3G_a$ and if $\eta < 0{\cdot}05$in, t $<$ 1in and $C > 1$in

$$\text{then } \delta > 8 \text{ and } n = \delta = \frac{2C^2 G_a}{E\, t\eta}$$

This indicates that the stress concentration can be reduced by

(i) Reducing the length of overlap.
(ii) Reducing the shear modulus of the glue.
(iii) Increasing the Young's modulus of the adherend.
(iv) Increasing the thickness of the adherend.
(v) Increasing the thickness of the glue line.

It is fairly obvious that (ii), (iii), and (iv) will increase the strength of the joint but it is not so clear in the case of (i) or (v).

(i) Reducing the length of overlap will increase the mean shear stress and so reduce the stress concentration factor at the end of the overlap. It is of course not possible to continue reducing the length below a certain minimum as a point is reached when even the mean shear stress in the joint is equal to the failing shear stress. This is illustrated by de Bruyne[9] when, with adherends 0·25in thick the joint strength was independent of overlap length down to an overlap of 2in, below this the joint strength is approximately linear with overlap length.

(v) Increasing the thickness of the glue line itself will decrease the shear stress on the glue line and although this will decrease the stress concentration factor it may not increase the joint strength. If however the strain rate is kept constant, increasing the glue line thickness does not decrease the joint strength as Bryant demonstrated in 1964.[10]

(b) *The Case of a Joint Between an Elastomeric Adherend and a rigid Adherend*

If one considers Volkersen's analysis as applied to a system where

$$\begin{aligned}
E_1 &= 1{\cdot}5 \times 10^3 \text{ lb/in}^2\\
E_2 &= 3 \quad \times 10^7 \text{ lb/in}^2\\
G_a &= 0{\cdot}5 \times 10^3 \text{ lb/in}^2\\
t_1 &= t_2 = 0{\cdot}5 \text{ in}\\
C &= 1 \text{ in}\\
\eta &= 0{\cdot}01 \text{ in}
\end{aligned}$$

The values of n at each of the two ends of the overlap can be calculated; these are of course different and are achieved by transposing E_1 and E_2.

These values are $8{\cdot}2 \times 10^{-4}$ and $11{\cdot}5 \sqrt{2}$.

This is very clearly demonstrable if you make up this joint and stretch it. It is very obvious that the stress concentration at the end of the steel adherend is very much greater than the stress concentration at the end of the rubber adherend.

Stress distribution due to stretching and bending

It is clear that the situation examined by Volkersen, whilst it indicates in a general way the importance of various parameters on the strength of a lap joint, is unsatisfactory in that it takes no account of the tearing stress set up in the adhesive as a result of the eccentricity of loading of the lap joint.

This is the system that was analysed by Goland and Reissner.[2] The joint they considered is again that shown in Fig. 8 where the Young's Modulus of the adhesive is E_a and that of the adherends is E. This analysis has since been studied by Sneddon[11] from whence most of this section has been developed.

The tensile stresses in the glue line σ_x and σ_y in the x and y directions as shown in Fig. 8 and the shear stress across a plane parallel to the glue line τ_{xy} were calculated as a function of the bending moment, K (OA/OB in Fig. 9) of the applied load. These stresses can only be evaluated however, if two approximations are used.

In the first approximation, where $\dfrac{\eta}{t} \ll \dfrac{E_a}{E}$ or more usefully where $\dfrac{\eta}{t} < \dfrac{E_a}{10E}$, the stresses in the glue line themselves are neglected and the

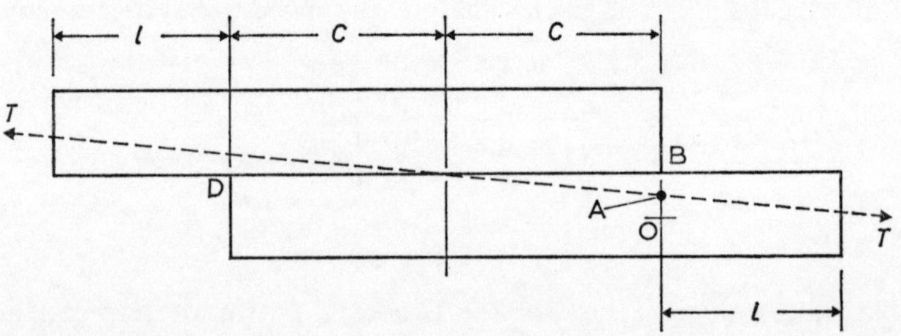

Fig. 9. Lap joint with thin glue line under load showing location of O, A and B

lap joint is considered as a uniform sheet of the shape of Fig. 9 with no glue line, where σ_x σ_y and τ_{xy} are calculated along the line BD of the sheet. This is the case of a relatively rigid adhesive in a thin glue line.

In the second approximation where $\frac{\eta}{t} \gg \frac{E_a}{E}$ or again more usefully where $\frac{\eta}{t} > \frac{10E_a}{E}$, the stresses in the adherends are ignored and σ_x σ_y and τ_{xy} are calculated along the mid plane of the adhesive layer. This can be applied to a relatively flexible adhesive in a thick glue line.

It is unfortunately not possible to calculate the stresses where $\frac{10E_a}{E} > \frac{\eta}{t} > \frac{E_a}{10E}$

(a) *Joints with a Rigid Adhesive in a Thin Glue Line*

The distribution of stress in a joint loaded in the x direction, i.e. where $K = 1$ is shown in Fig. 10. From this it can be seen that whilst the tearing stress (σ_y) and the longitudinal tensile stress (σ_x) are both high at the edge of the joint, the shear stress is zero. It is possible to compound the three stresses at a point such as ω_o where the stresses σ_{xo} and σ_{yo} and τ_{xyo} give a shear stress of τ_o across some plane at ω_o along the joint but at an angle to the glue line. If this is calculated at all points along the glue line a curve of τ_o against ω_o can be drawn and this will itself have a maximum (τ_o) max. For a given system there will thus be a unique (τ_o) max for each K value.

Greenwood, Boag and McLaren[12] examined a joint between steel adherends bonded with an epoxy resin where

$$\begin{aligned} E_1 = E_2 &= 3 \times 10^7 \text{ lb/in}^2 \\ E_a &= 4{\cdot}5 \times 10^3 \text{ lb/in}^2 \\ t_1 &= t_2 = 0{\cdot}5 \text{ in} \\ C &= 1{\cdot}5 \text{ in} \\ \eta &= 0{\cdot}004 \text{ in} \end{aligned}$$

In this joint $\frac{\eta}{t} = 0{\cdot}008$ and $\frac{E_a}{E} = 0{\cdot}015$. Whilst $\frac{\eta}{t}$ is not less than $\frac{E_a}{10E}$ it is certainly less than $\frac{E_a}{E}$ and it was assumed that the first approximation applies and the relationship between (τ_o) max and K is shown in Fig. 11.

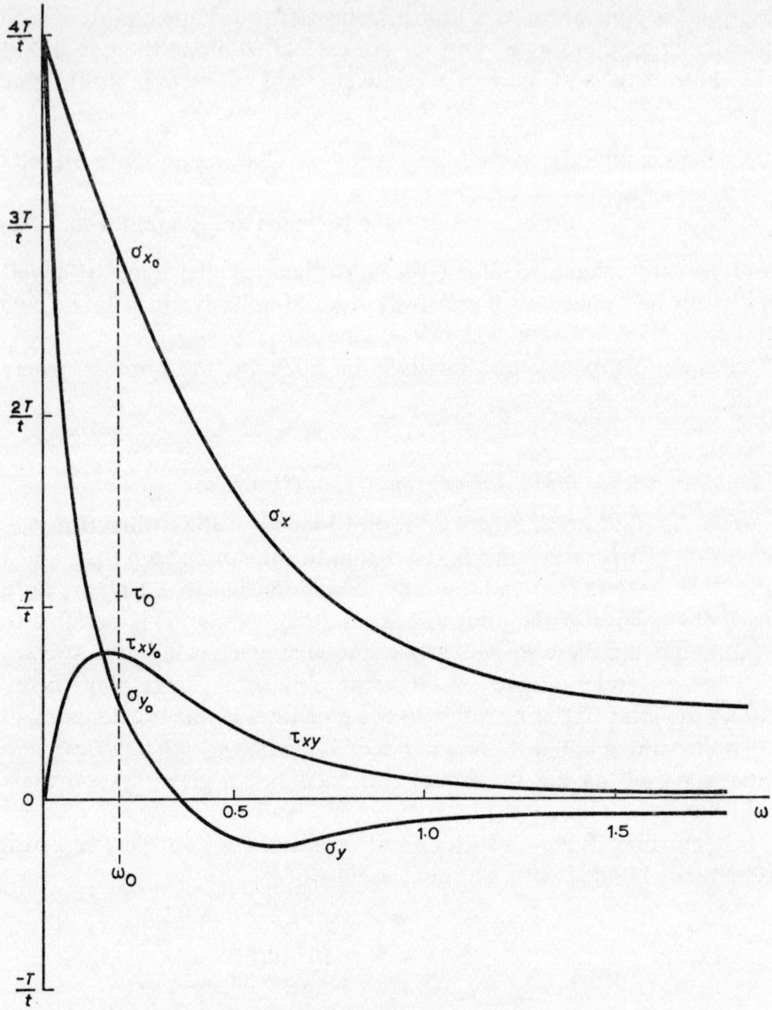

Fig. 10. Stress distribution along the shear plane of a joint with a rigid adhesive and K=1 (ω is distance from joint edge/sheet thickness)

Fig. 12 shows the observed values of $\frac{T}{t}$ where T is the failing load of the joint per inch width and t the thickness of the adherend, plotted against the K value of the joint. From this curve the advantages of loading the joint in such a way that the K value of the joint lies between $-\frac{1}{3}$ and $-\frac{2}{3}$ is clearly demonstrated.

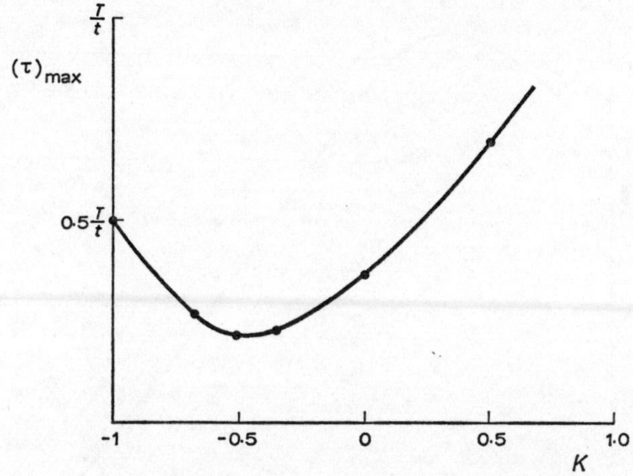

Fig. 11. Relationship between $(\tau_0)_{\max}$ and K

From Figs. 11 and 12 we have the values of (τ_0) max and $\frac{T}{t}$ for each K value and from this the values of (τ_0) max can be calculated as follows.

TABLE 3

K value	(τ_0) max (lb/in²)
-1	3 700
$-\frac{2}{3}$	3 940
$-\frac{1}{3}$	3 240
$\frac{2}{3}$	3 000
$+\frac{1}{3}$	4 350
$+\frac{2}{3}$	6 500 *
Mean	3 650

*As the adherend bent during this test this figure is ignored.

The failing shear stress of the bulk adhesive was measured as 3 212 lb/in² which is comfortably near to the (τ_0) max measured in the joints. This suggests that when the failing shear stress in the adhesive is reached the joint fails and this shear stress can be related to the failing load of the joint by its physical properties and its geometry.

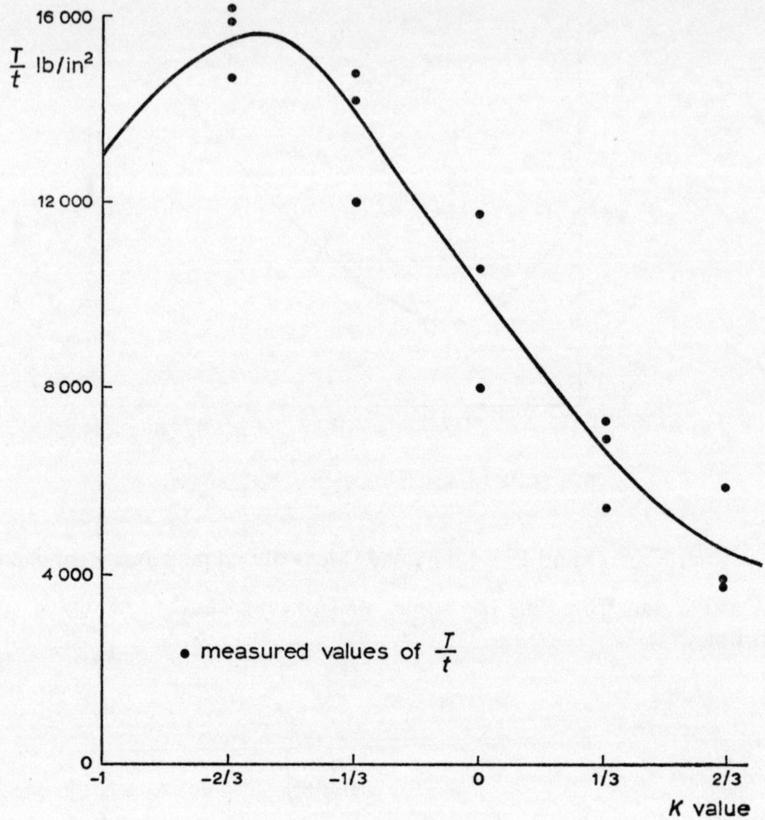

Fig. 12. Observed values of $\frac{T}{t}$ per inch width of the joint plotted against the K value of the joint

(b) *Joints with a relatively Flexible Adhesive in a Thick Glue Line*

Greenwood, Boag and McLaren[12] repeated their work on using a flexible adhesive, a silicone rubber between their steel adherends. Here the joints had the following properties;

$$E_1 = E_2 = 3 \times 10^7 \text{ lb/in}^2$$
$$E_a = 450 \text{ lb/in}^2$$
$$t_1 = t_2 = 0{\cdot}5 \text{ in}$$
$$C = 1{\cdot}5 \text{ in}$$
$$\eta = 0{\cdot}011 \text{ in}$$

Here $\frac{\eta}{t} = 0{\cdot}022$ and $\frac{E_a}{E} = 1{\cdot}5 \times 10^{-5}$ and hence as $\frac{\eta}{t} \gg \frac{E_a}{E}$ the second approximation can be applied.

Again using Sneddon's analysis[11] the value of (τ_0) max is dependent upon a parameter $\frac{\beta C}{t}$ where

$$\beta = \sqrt{8G_a\ (1-\omega^2)\ t/E\eta}$$

where G_a is the shear modulus of the adhesive, 150 lb/in² and ω its Poissons ratio, 0·5.

In the joint considered $\frac{\beta C}{t}$ has a value of 0·1 and at this value according to Sneddon (τ_0) max should be independent of the K value and this was shown to be the case in Table 4.

TABLE 4

K value	Failing load (lb/in²)
-1	443
$-\frac{2}{3}$	427
$-\frac{1}{3}$	418
0	365
$+\frac{1}{3}$	442
$+\frac{2}{3}$	453

In order to take advantage of the influence of the bending moment one must increase the value of $\frac{\beta C}{t}$ and this can be done by either increasing the shear modulus of the glue, increasing the thickness of the adherend, reducing the Young's modulus of the adherend or reducing the glue line thickness. This was confirmed by using the same epoxy resin adhesive in ASTM : D 1002[13] test specimens.

Here

$$E_1 = E_2 = 3 \times 10^7\ \text{lb/in}^2$$
$$E_a = 4{\cdot}5 \times 10^5\ \text{lb/in}^2$$
$$t_1 = t_2 = 0{\cdot}625\ \text{in}$$
$$C = 0{\cdot}25\ \text{in}$$
$$\eta = 0{\cdot}004\ \text{in}$$

During loading the K factor was observed as $+0{\cdot}5$

$$\frac{\eta}{t} = 0{\cdot}06 \text{ and } \frac{E_a}{E} = 0{\cdot}015$$

and thus, as $\frac{\eta}{t} > \frac{E_a}{E}$ the second approximation most nearly applies.

Here $\frac{\beta C}{t} = 2{\cdot}7$ and according to Sneddon[11] if we take the value of (τ_0) max we calculated earlier as 3650 lb/in²

$$T = (\tau_0)\max \times C = 3650 \times 0{\cdot}25 \text{ lb}$$
$$= 912 \text{ lb}$$

The observed values had a mean of 958 lb/in².

Conclusion

The strength of a lap joint can be influenced by the geometry of the joint and the physical properties of the components of the joint. These physical properties can be varied to vary the strength of the joint. The joint strength can be increased if a small negative bending moment can be imposed upon it. Assuming all joint failures are cohesive in the adhesive, the strength of a joint can be forecast from an examination of the physical properties of the components and the geometry of the system.

References

1. Volkersen, O., *Luftfahrtforsch*, **15**, 41, 1938.
2. Goland, M. and Reissner, E., *J. Appl. Mech.*, **2**, A–17, 1944.
3. Plantema, F. J., *De Schuifspanning in eme lijmnaad*, Report M1181, Nat Luchtvaartlaboratorium, Amsterdam, 1949.
4. Cornell, R. W., *J. Appl. Mech.*, **75**, 355, 1953.
5. Lubkin, J. L. and Reissner, E., *J. Appl. Mech.*, **78**, 1213, 1956.
6. Lubkin, J. L., *J. Appl. Mech.*, **24**, 255, 1957.
7. Mylonas, C and de Bruyne, N. A., *Adhesion and Adhesives*, Ed. de Bruyne, N. A. and Houwink, R., Elsevier, Amsterdam.
8. McLaren, A. S. and McInnes, I., *Brit. J. Appl. Phys.*, **9**, 72, 1958.
9. De Bruyne, N. A., *Aircraft Eng.*, **16**, 115, 140, 1944.
10. Bryant, R. W., *Nature*, **202**, (4937), 1087, 1964.
11. Sneddon, I. N., *Adhesion*, Ed. Eley, D. D., p. 219, Oxford University Press, 1961.
12. Greenwood, L., Boag, T. R. and McLaren, A. S., *Adhesion: Fundamentals and Practice*, Ministry of Technology, Maclaren, London, 1969.
13. American Society for Testing Materials ASTM D: 1002.

The Assessment of Structural Bonds by Destructive Methods

P. E. GALLANT, B.Sc., A.R.C.S., A.R.I.C., and C. S. SWAFFER†*

IMPERIAL METAL INDUSTRIES (KYNOCH) LIMITED

Introductory

Structural bonding implies the use of adhesives in an engineering application and the adhesive technologist must be able to satisfy the design engineer that the proposed bonded joint is capable of withstanding the stresses to be transmitted by it. These stresses will be applied either in shear or tension, or in combination, but design preference is given to joints utilising adhesives in shear. It has long been customary in the metallurgical field for prime design data to be obtained from the preparation and testing to failure of suitable test pieces, but while this method is satisfactory for homogeneous materials, an adhesively bonded joint is a composite and as such requires a different approach. The design of a bonded joint has a profound effect on the performance of the adhesive, and when test figures obtained from destructive tests are quoted it is also necessary to quote the test method, the nature of the adherends, and the type of surface preparation employed. The main benefit derived from destructive testing is the direct comparison of one adhesive with another when tested under the same conditions, but unless the stress distribution in the test joint is comparable with that to be met in service, the direct application of laboratory data is extremely difficult. It is therefore important to select carefully the tests to be employed to obtain design data before a satisfactory design can be produced.

The successful application of adhesives to structural bonding depends on three clearly defined conditions, (i) the physical properties of the adherends, (ii) the physical properties of the adhesive and (iii) the properties of the adhesive-adherend system under the conditions in which the structure is to operate. Of these, information

* Now at I.C.I. Fibres Ltd, Pontypool

† Now at Formica Ltd, North Shields

on the first can usually be obtained by reference to the literature, but for the second, laboratory investigation may often be necessary. The third, and most important, can only be determined once the adherend system has been defined, and the correct design of a bonded joint depends on close liaison between the adhesive technologist and the design engineer.

In this paper the most important condition, the marriage of the adhesive and adherends, is being considered, with particular reference to the test methods which may be used, and their application to the development of the Imperial Metal Industries (Kynoch) Limited process for the production of lightweight pressure vessels by adhesive bonding.

Destructive test methods

By definition, the term structural bond is used where an adhesive transmits a load between one adherend and another, and is generally taken to apply to metal-to-metal, wood-to-wood, or wood-to-metal bonded joints. Although this paper is mainly concerned with metal-to-metal bonds, it is also important that wood-to-wood applications are considered, for there are many instances of adhesives being successfully employed in the fabrication of wooden structures, e.g. the use of laminated wood beams in buildings. The methods that may be adopted for destructive testing will now be considered in two parts. Firstly, the generally accepted standard methods and secondly, the methods developed and employed at IMI.

Part 1: Standard test methods

(*a*) *Wood*

The established destructive tests for wood adhesives in the UK are detailed in the relevant British Standard Specifications 1203, 1455, 1204 and 3544. In BS 1203 a three ply test panel is cut into strips one inch wide, and further cut to form overlap joints (Fig. 13). A standard wood (beech) is used, and the tensile strength of the joints is determined dry, and after immersion in water at a given temperature for set periods of time. BS 1455 describes the well known knife test. A specially shaped chisel/knife is used to prise the outermost lamina from the test panel, and the result is compared with a set of standard panels or photographs. The better the adhesive the more difficult it is to separate the laminae.

BS 1204 is written in two parts. Part 1 concerns gap filling adhe-

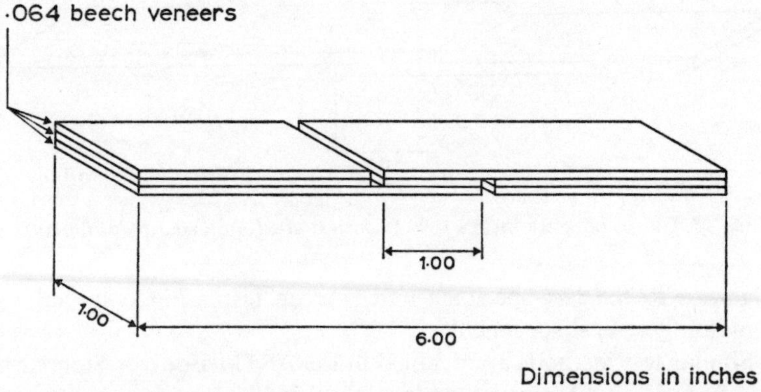

Fig. 13. Plywood lap joint BS 1203 (*Courtesy British Standards Institution*)

sives, and again a lap joint is used as the test specimen. This is prepared by using two one inch wide strips of beech, and bonding them together with a 0·050 in spacer between them to form the glue line (Fig. 14). Part 2 concerns close contact adhesives, where two one-eighth inch thick wood adherends are bonded together to form a conventional lap joint one inch wide, with one inch overlap (Fig. 15). In both parts the tests are conducted under the same conditions as for BS 1203. In these specifications it is hoped to achieve failure in the adherend, but an assessment of the percentage of bond failure aids the interpretation of results.

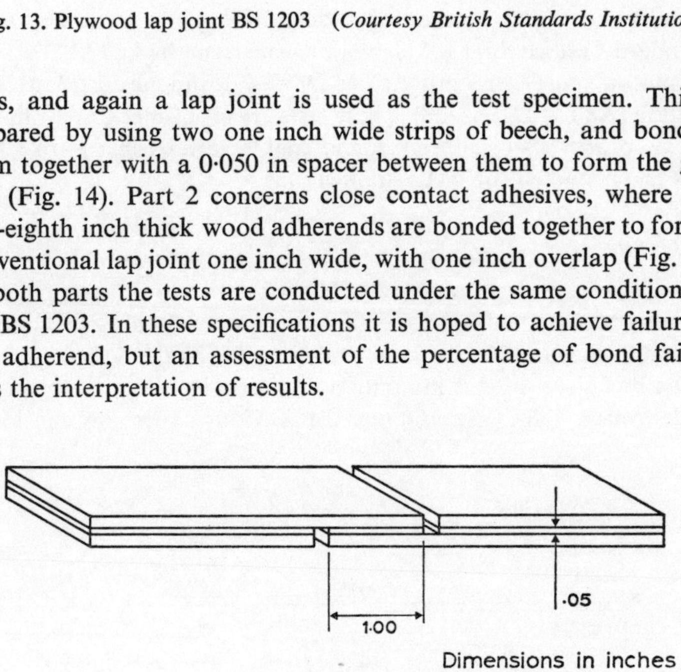

Fig. 14. Gap joint BS 1204 (*Courtesy British Standards Institution*)

The evaluation of polyvinyl acetate wood adhesives is covered by BS 3544 which specifies the use of a double lap joint test piece. This specification prescribes a three point bend test with the load applied

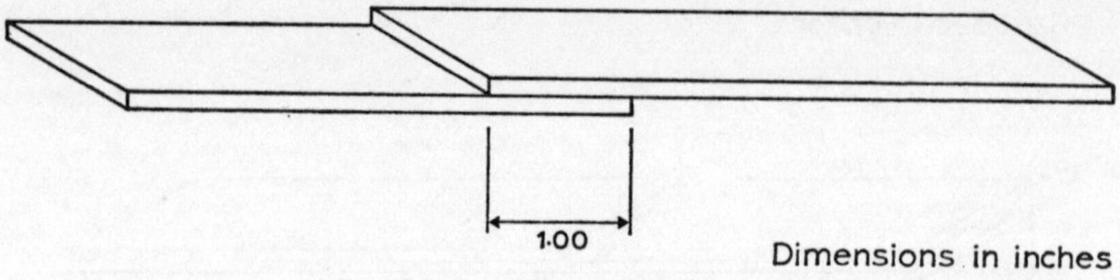

Fig. 15. Close contact joint BS 1204 (*Courtesy British Standards Institution*)

edgewise to the test specimen and a creep test using a special rig containing several specimens.

Similar test methods are detailed in the ASTM Book of Standards and Federal Test Method Standard No. 175. In these two documents will also be found methods for determining the impact strength (Standard Test Method 1051), block shear strength (ASTM D905-49), the 'button' tensile strength (ASTM D897-49), and the cross lap tensile strength (ASTM D1344–57). These tests are all of merit in evaluating specific properties of adhesives, and enable various species of adherends to be studied for given applications.

(*b*) *Metal*

There are a number of well established destructive tests applicable to metal-to-metal adhesives available to the technologist, although the results may not be in a form readily usable by the design engineer.

The test piece most commonly used to assess shear strength is the single overlap joint (Fig. 16), and the previous paper[1] has considered

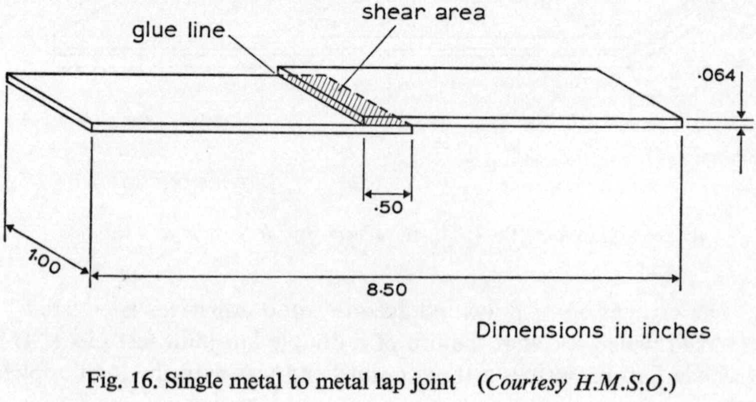

Fig. 16. Single metal to metal lap joint (*Courtesy H.M.S.O.*)

the stresses therein. It is simple to manufacture and test, and is applicable to a wide range of adherends. Standard methods of preparation are specified in DTD 861 where joints are prepared individually by the assembly of two component pieces (Fig. 17), and in

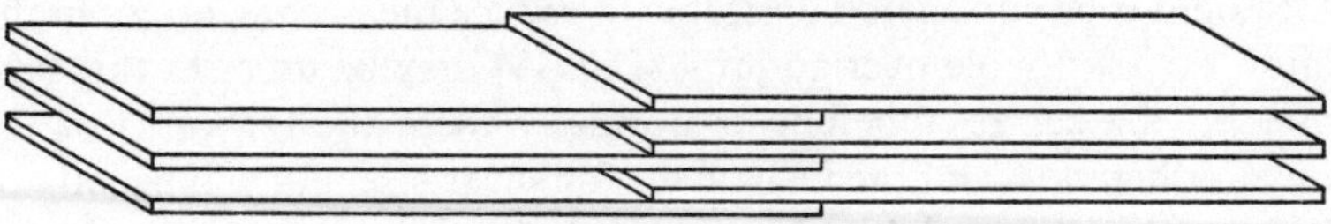

Fig. 17. Individually assembled lap joint

DTD 5577 where two panels (plain or slotted) are bonded together with a given overlap (Fig. 18) and subsequently machined to make individual test pieces. The former method may be more laborious, but machining of the bonded joint, a process that can easily lead to damage to the bond, is avoided. However, the latter method agrees with those detailed in US Military Specification Mil.–A–5090D and ASTM D1002. ASTM D1002 is more detailed than DTD 5577 and contains several points of warning, particularly regarding the selection of adherend gauge thickness to avoid exceeding the 0·1 per cent proof stress of the material during test, and precautions to be taken in cutting test pieces from the bonded panel. It is perhaps pertinent to point out that whilst it may be attractive to make panels of more than one specimen, the use of very high strength adherends with their associated machining problems, enforces manufacture of individual

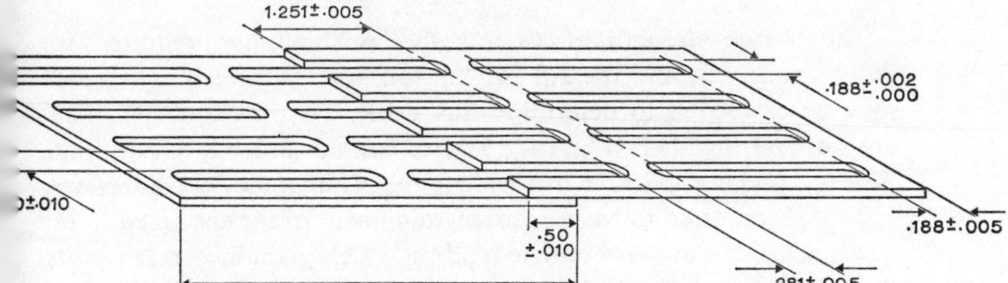

Fig. 18. Slotted panel lap joint assembly (*Courtesy H.M.S.O.*)

test pieces on economic grounds. In spite of its wide use, this test does not reproduce conditions likely to be met in practice, but does enable the evaluation of adhesives and adherends to be made on a comparative basis, giving quantitative data rather than a series of subjective observations. The test piece may also be used to determine joint behaviour under long-term storage conditions involving extremes of temperature and humidity.

Should it be considered desirable to reduce the peeling forces in the joint, then a double overlap joint (Fig. 19) may be used. In this type of joint the centre adherend is normally twice the thickness of the outer adherends, and the test values for shear strength will usually be higher than those obtained with single lap joints, more nearly approaching the pure shear strength of the adhesive. Use of this specimen is limited by the need for a higher capacity test machine and for more adherends of differing thicknesses.

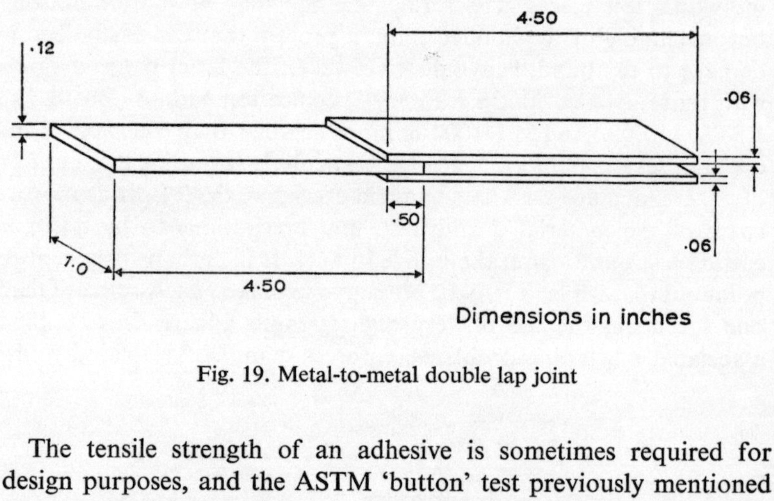

Fig. 19. Metal-to-metal double lap joint

The tensile strength of an adhesive is sometimes required for design purposes, and the ASTM 'button' test previously mentioned may be employed to determine this value. Two 'button' specimens are bonded together (Fig. 20), and separated under a tensile load. This method is applicable to a wide range of adherends and adhesives, but care is required to ensure correct alignment of the specimen in the test machine to avoid eccentric loading which gives incorrect results. An alternative method which can be used conveniently with adhesives cured under low pressures, is the butt joint tensile test referred to in ASTM D2095–62T (Fig. 21). Here, two bars half inch square are bonded together end to end, and subsequently pulled apart using self aligning test grips.

Plate 4 CIBA peel test

PLATE 5 'T' peel test

PLATE 6 Climbing drum peel test (*Courtesy American Society for Testing Materials*)

PLATE 7 IMI lap joint assembly jig

PLATE 8 Napkin ring shear test rig

Plate 9 Napkin ring test piece (exploded view)

Plate 10 Bonded washer test specimen

PLATE 11 Bonded washer torsion shear test rig

PLATE 12 Tube and rod test rig

PLATE 13 IMI Strip Laminate Pressure Vessel after pressurizing to destruction showing typical failure in hoop

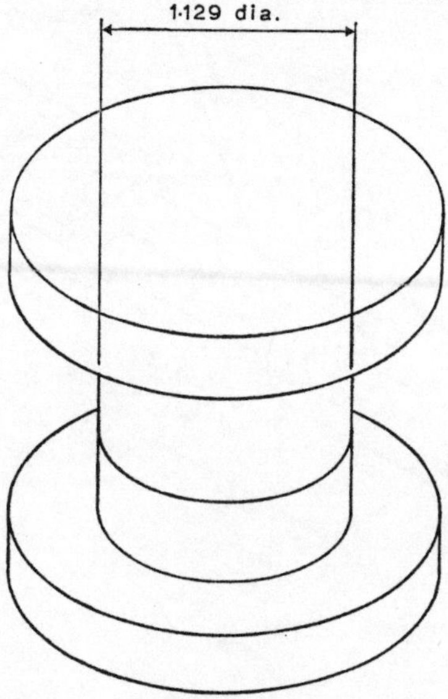

Fig. 20. ASTM button tensile test (*Courtesy American Society for Testing Materials*)

The pure shear strength and the shear modulus of an adhesive are the two characteristics of most use to the design engineer, but are not the easiest to determine. Over the past fifteen years a number of methods have been accepted as suitable, most of which employ a form of torsional testing. An early method developed in Canada known as the Napkin Ring test (Fig. 22) consisted of bonding the rims of two cups together, and twisting them to failure. This method is rather difficult to apply, since it is not easy to achieve a controlled glue line thickness, and the recommended test rig does not prevent tensile stresses being applied to the test piece.

In a variation of this method described in the US Forest Products Laboratory Report 1851, a multiplicity of washers are bonded together (Fig. 23), enabling the shear modulus to be determined by taking advantage of the extra movement gained from multiple glue lines. This has now been developed further, and ASTM E229–63 describes a torsion shear method where single or multiple glue lines are used (Fig. 24). As with the Napkin Ring test, care is required to

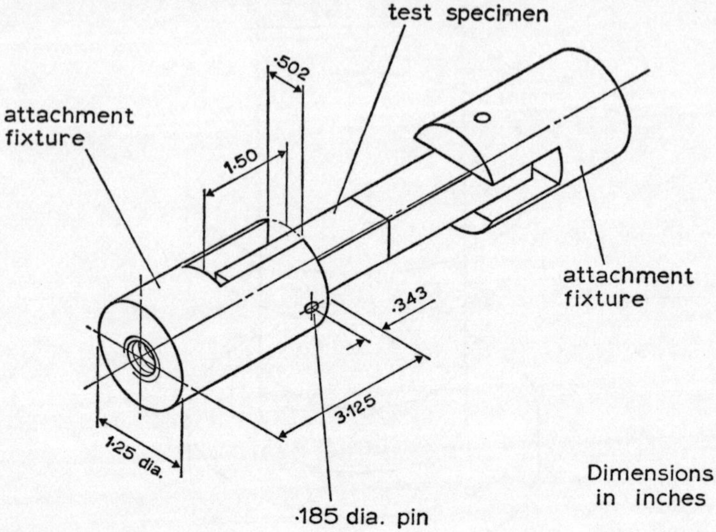

Fig. 21. ASTM butt joint test (*Courtesy American Society for Testing Materials*)

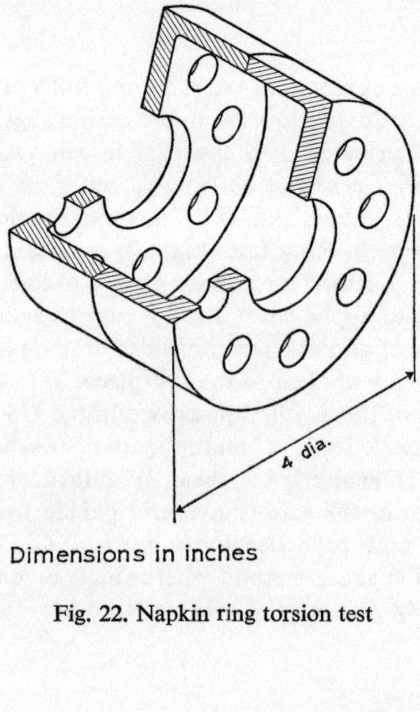

Fig. 22. Napkin ring torsion test

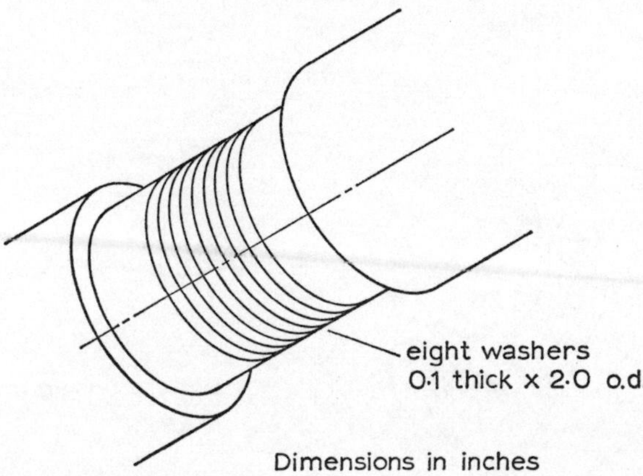

Fig. 23. Multiple washer torsion test (*Courtesy US Department of Agriculture*)

produce an accurately aligned test specimen, and a very sensitive extensometer is needed to determine the shear modulus.

Another method which prevents the introduction of tensile stresses was originally developed by the Raytheon Corp. in the USA and uses washers bonded to end plates (Fig. 25), with subsequent testing to failure by the application of a torsional load. This method is probably the cheapest to operate, and, as will be shown later, gives results comparable with those achieved with the Napkin Ring test.

The peel strength of adhesives is frequently of value in the design of bonded structures, and a number of methods have been established to evaluate this property. The CIBA (ARL) Ltd peel test (Plate 4) consists of two flat strips bonded together and wrapped round a six inch diameter drum. The inner strip is secured to the drum, and the outer strip pulled away from it, the applied load being recorded graphically. Another type of peel test is the 'T' peel test (Plate 5) in which two strips are bonded together along 75 per cent of their length, the unbonded ends bent through 90 degrees, and placed in the jaws of a tensile test machine to separate them. This test is most suitable for very high peel strength adhesives, and the results are again recorded graphically.

The generally accepted method at present is the climbing drum test (Plate 6) specified in ASTM D1781–62, and DTD 5577. It is intended for the evaluation of adhesives for bonding face sheets to honeycomb cores, but it can also be used with metal-to-metal bonded sheets.

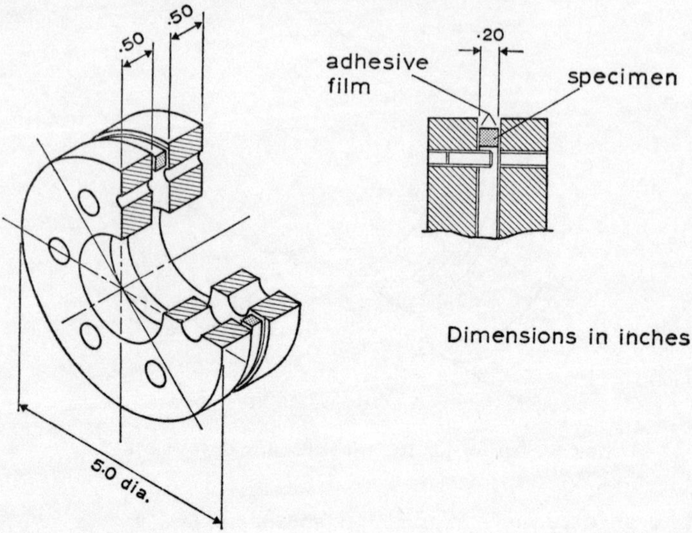

Fig. 24. ASTM torsion test (*Courtesy American Society for Testing Materials*)

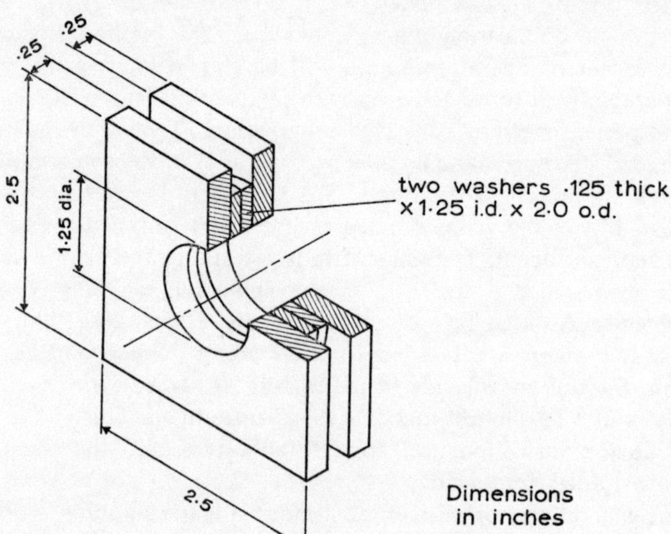

Fig. 25. Raytheon bonded washer torsion test

These three methods are specialised tests, and the results are particularly influenced by the physical properties of the adherend. Using given adherends comparative values for a range of adhesives can be obtained.

It is appreciated that this survey has described only the principal methods of evaluating structural adhesives, and that ease of handling, cost, size of structure and end use all have to be considered when deciding which method to use in a particular instance.

It is equally important to realise that the results obtained can only be used as design data when correlated with the performance of a real structure.

Part 2: Application of destructive tests to a specific problem

When a bonded structure is designed, laboratory data obtained by the methods just described cannot always be used directly, since these tests rarely reproduce the stress system encountered in practice. For example, in the work about to be described, the stresses in the adhesive bonds are biaxial. The aim was to produce very high specific strength pressure vessels, using steels with a tensile strength in excess of 100 tons/in^2 and which, because of their metallurgical properties, could only be effectively used in a bonded structure. This work resulted in the IMI strip laminate process, in which a tubular section is formed by helically winding layers of flat metal strip onto a removable mandrel, bonding them together, and completing the vessel by bonding the tubes so formed, either externally or internally, to end fitments.

The design of a strip laminate pressure vessel follows conventional practice for the determination of wall thickness, but the overlap length of the end fitments, and the strip overlap pattern for optimum performance, need to be carefully considered in relation to the properties of the adhesives employed. Examination of a cross section of a typical tube wall structure (Fig. 26), shows that the principal stresses in the adhesive are in shear, and because with metal adherends the preferred adhesives are based on epoxy resins, the majority of work associated with this process has been concerned with this system.

In selecting the tube wall adhesive it was essential that it should be capable of transmitting shear stresses as great as those to be carried by the end fitment bond, and once this objective has been obtained the construction of a tube wall (number of layers and strip width)

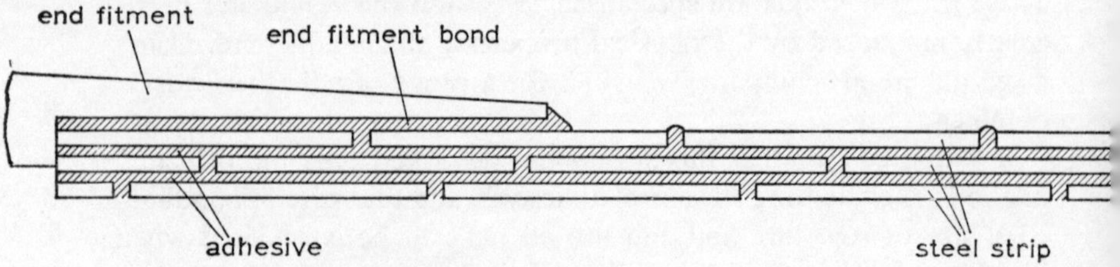

Fig. 26. Section through tube and end fitment junction

and the overlap pattern of successive layers can be established. In general, a minimum of three layers and a maximum strip thickness of 0·015 in are employed. The adhesive should also be capable of meeting the performance requirements in respect of operating temperatures and resistance to climatic extremes.

The adhesive used to attach end fitments to the tubular section must be capable of carrying high shear stresses so that an efficient design may be achieved, since the higher the design stress in the adhesive, the shorter the overlap length required to meet a given pressure requirement. In addition to possessing a very high shear strength, the adhesive in this instance must be capable of withstanding the peeling forces at the end fitment to tube wall junction, particularly with internal attachment, and must also meet the same performance requirements as the tube wall adhesive.

The evaluation of adhesives to satisfy these conditions has utilised both laboratory and model structure tests. The principal use of laboratory tests has been the comparative assessment of alternative formulations, and the establishment of optimum cure conditions for thermosetting materials, while design data has been obtained from hydrostatic burst tests of scale structures.

In the early stages of development a standard method of surface preparation was established, which involved vapour degreasing, shot blasting with an 85 mesh alumina grit, followed by a further degrease. This method has been applied in all laboratory tests as well as in the actual process. The principal adherend used in development work was high strength steel, although other metals have also been evaluated for specific applications.

Laboratory test methods

Initial evaluation of potential adhesive systems employed a lap joint test using adherends with a minimum tensile strength of 100 ton/in^2. To suit the use of these very high strength adherends the accepted methods were adapted by punching a hole in one end (Fig. 27) of a standard test piece 0·065 in thick and 1·18 in wide. The choicc of width was made purely on availability of material.

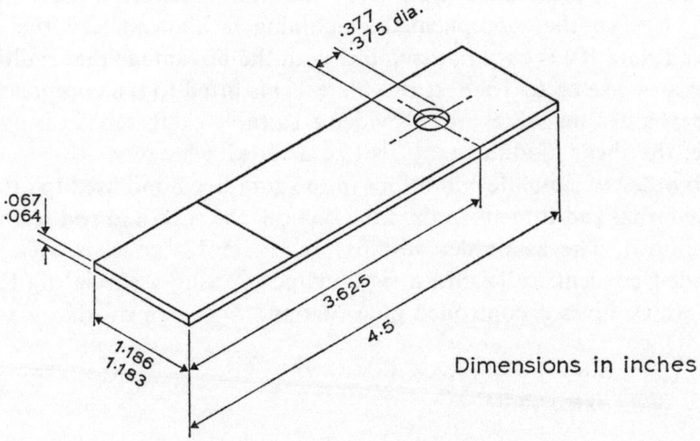

Fig. 27. IMI lap joint test piece

A number of individual test pieces is assembled in a jig (Plate 7) and the hole in each test piece used to maintain correct overlap of the bonded joint. Correct glue line thickness and longitudinal alignment are obtained by the use of spacers, and the assembly is compressed by a spring during the cure cycle. As the adherends are of such high strength it has been found possible to make double lap joints from three standard test pieces, without risk of exceeding the 0·1 per cent proof stress of the steel during test, thus effecting considerable savings in manufacturing costs; the same assembly jig can also be used to prepare double lap joints. Test joints are pulled in a standard tensile test machine using the hole in each component piece to assist mounting in the test grips.

In the early stages of development attempts were made to determine the peel strength of the adhesives most suited to this process, but little success was achieved with either the CIBA peel test or the

'T' peel test, because of the stiffness of the adherends. Recently, however, the climbing drum peel test has been under evaluation for another application of these adhesives.

The pure shear strength of adhesives is of much interest in this process and two methods have been evaluated. The first was the Napkin Ring method shown in Plate 8 and attention was given to the design of the test cups and spacers to ensure that a uniform glue line could be achieved (Plate 9). The second test method used was the bonded washer technique. Individual components are made from flat plate so that complicated machining is avoided and the test piece (Plate 10) is easy to assemble, with the advantage that multiple glue lines are used. The test rig (Plate 11) is fitted to the compression side of a test machine, and provided a suitable extensometer is available, the shear modulus may also be determined.

In order to simulate conditions in the adhesive bond used to attach an external end fitment to the tube section, the tube and rod test was developed. The assembled test fixture (Plate 12) consists of a rod bonded concentrically into a rigid cylinder, using a special jig (Fig. 28) which gives a controlled glue line and a known overlap length.

Laboratory test results

TABLE 5

Lap Shear Strengths for Different Adhesives (Steel Adherends). (All figures are the mean of 6 individuals)

Adhesive	Single overlap (lb/in^2)	Double overlap (lb/in^2)
CIBA BSL 309	2 740	4 660
CIBA BSL 308	6 300	7 870
Bloomingdale FM 77	4 540	5 590
Bloomingdale FM 1044	6 160	6 630
Bloomingdale HT 424	4 600	5 450
Bloomingdale FM 1000	6 600	6 810
Araldite AV 100 + HV 100	4 580	4 250
Araldite X33/1202	5 600	6 400
Araldite AZ 15	6 550	7 600
Bloomingdale FM 123	5 960	6 610
CIBA ARA 2800	3 480	5 450
Araldite MY 753 + HT 972	4 630	5 950

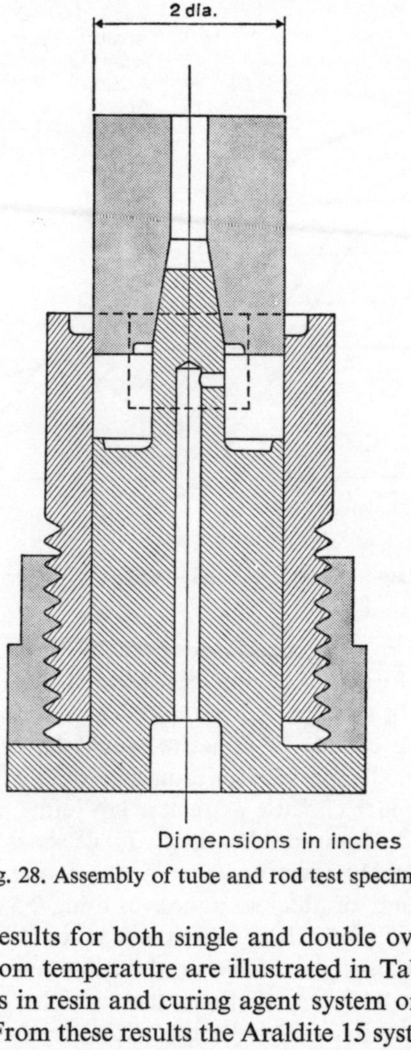

Fig. 28. Assembly of tube and rod test specimen

Typical test results for both single and double overlap joint tests conducted at room temperature are illustrated in Table 5, where the effect of changes in resin and curing agent system on joint strength are compared. From these results the Araldite 15 system was selected for use as the tube wall adhesive, and Araldite X33/1202 adhesive for end fitment attachment. At the time of testing these two formulations gave the highest lap shear strengths, and were readily applicable to the process. The adhesive performance at various temperatures has also been studied using double lap joints, with the results given in Fig. 29. Here the temperature has been recorded by attaching thermocouples to the bonded joints, and it can be seen that there is a

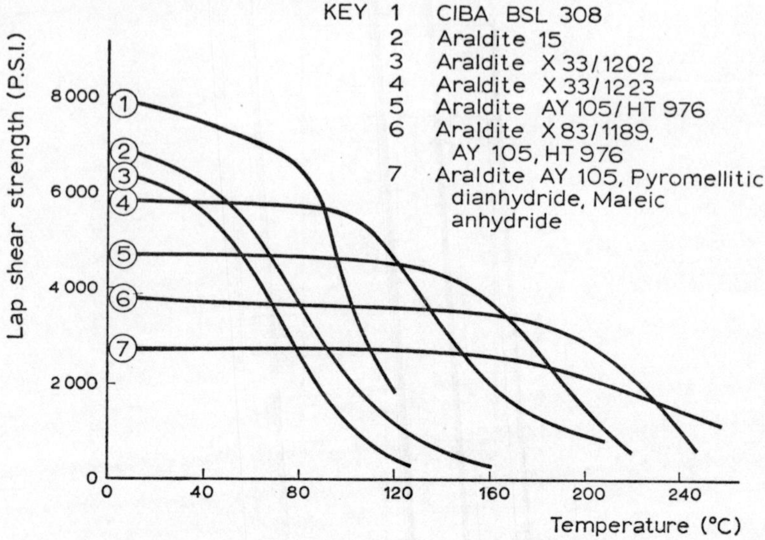

Fig. 29. Lap joint shear strength—variation with temperature

general loss of room temperature strength for any gain in high temperature performance. It will be observed that the two selected adhesives do not possess outstanding elevated temperature performance, and a change of adhesive would be necessary to meet such conditions. To provide information on the performance of bonded joints under climatic extremes, lap joints have been stored under Intensified Standard Alternating Trials series B (Appendix A) conditions, a weekly cycle covering a range of temperatures and humidities. A study of different adhesives using 0·5 inch single overlap joints is found in Fig. 30, where it is seen that some formulations perform better than others. However, the importance of the geometry of the final structure when examining these results is shown in Fig. 31 where varying overlaps have been tested using a single adhesive. The lap joint test has also been used to assess the effect of adding fillers to particular adhesive formulations and this is shown in Fig. 32 where certain fillers improve the average shear stress achieved. The effect of changing the adherend has also been studied using the lap joint test, with the results shown in Table 6. These demonstrate that regardless of the choice of adherends it is possible to achieve very high lap joint shear stresses.

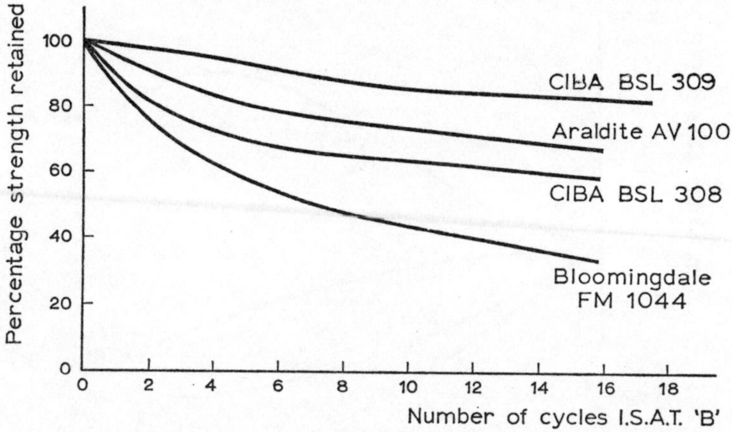

Fig. 30. ISAT 'B' storage tests of single lap joints using different adhesives

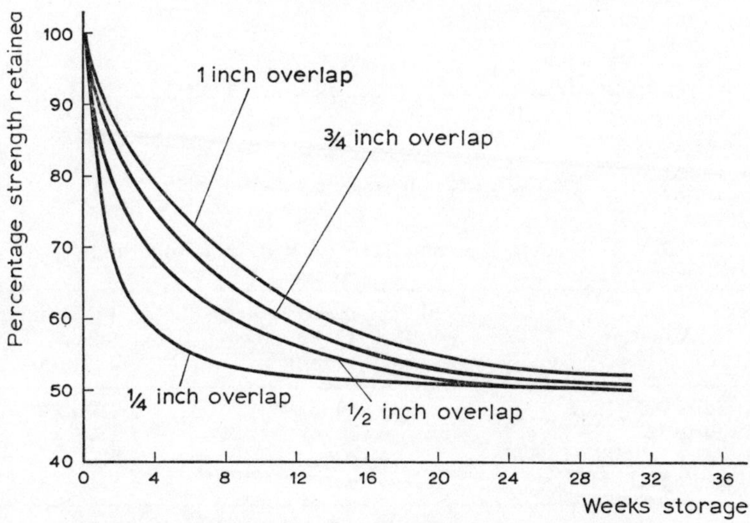

Fig. 31. ISAT storage trials of single lap joints of various overlaps

The results of torsion shear testing shown in Table 7 using both the Napkin Ring and the bonded washer test demonstrate that very high shear stresses can be achieved with modern adhesives when used in conjunction with high strength adherends. It is particularly noteworthy that the Araldite 15 adhesive will give pure shear stresses in excess of 10 000 lb/in^2.

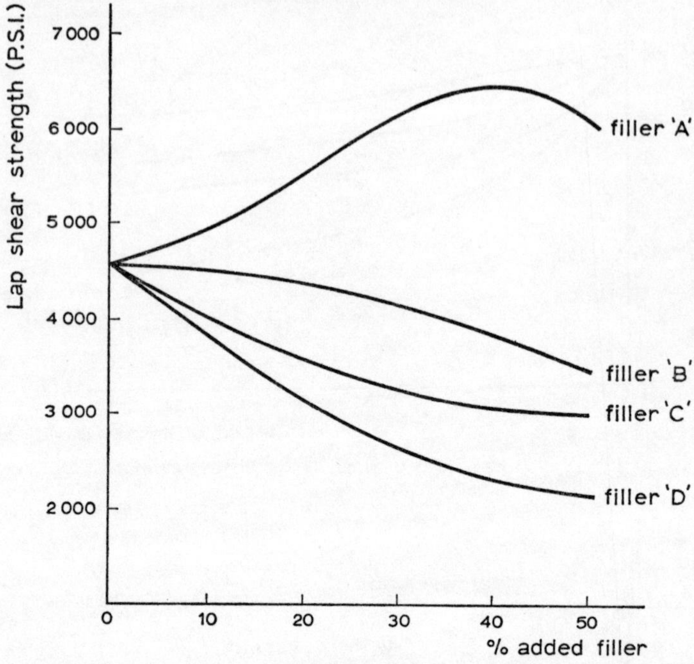

Fig. 32. Effect of fillers on lap joint strength

TABLE 6 Double Lap Joint Tests with various Adherends

Adhesive	0·95% carbon steel BS 1449, En 44D (100 ton/in^2)	6 AL 4V Titanium alloy (60 ton/in^2)
Araldite X33/1202	5 910 lb/in^2	5 270 lb/in^2
Araldite 15	6 930	7 280
Araldite AY 105/HT 976	4 130	4 350
Bloomingdale HT424	5 740	4 945

Adhesive	Stainless steel (100 ton/in^2)	Light alloy (35 ton/in^2)
Araldite X33/1202	6 150 lb/in^2	Metal failure
Araldite 15	7 630	
Araldite AY 105/HT 976	5 170	
Bloomingdale HT424	5 750	

TABLE 7

Torsion Shear Test Results

Adhesive	Napkin ring test (lb/in²)	Bonded washer test (lb/in²)
Araldite 15	9 225	10 340
Araldite 15	8 490	10 800
Araldite 15	8 830	10 540
Araldite 15	8 290	10 720
Araldite X33/1202	8 050	8 500
Araldite X33/1202	7 460	8 500
Araldite X33/1202	8 050	8 370
Araldite X33/1202	7 360	8 500
Araldite MY 753/HT 972	8 240	8 760
Araldite MY 753/HT 972	8 590	9 090
Araldite MY 753/HT 972	8 490	8 380
Araldite MY 753/HT 972	8 930	7 670

Small scale pressure vessel tests (6·5 *in diameter*)

It will be appreciated that the strip laminate technique offers considerable freedom in respect of the number of layers and strip widths employed to satisfy a particular pressure vessel design. The overlap length used to attach the end fitments will also vary according to the pressure requirement to be met. In the search for reliable design information, many tests have been conducted on 6·5 in dia pressure vessels to study specific points of design.

The strength of the tube wall adhesive, Araldite 15, has been investigated in a two layer tube, using comparatively thick strip (0·012 in) and varying the amount by which the second layer overlaps the first. The mean adhesive shear stresses at bursting obtained in these tests are shown in Table 8. The figures very closely approach those obtained in the double overlap shear tests, but are considerably lower than those obtained in torsion. It is also evident that when the overlap was increased to 0·6 in failure was attributable to overstressing of the steel, but with smaller overlaps shear stresses in excess of 6 000 lb/in² in the adhesive bond have been achieved.

Attachment of end fitments presents a different problem since the pressures required to produce end loads sufficient to exert mean shear stresses in the adhesive bond in excess of 5 000 lb/in² with an

TABLE 8 Hydrostatic Burst Tests of 2 Layer Test Vessels

Overlap length (in)	Failing pressure (lb/in²)	Hoop stress in steel strip at failure (lb/in²)	Adhesive shear stress at failure (lb/in²)	Mode of failure
0·2	850	115 000	6 930	Adhesive in shear
0·2	830	112 000	6 760	Adhesive in shear
0·3	1 210	163 000	6 570	Adhesive in shear
0·3	1 250	169 000	6 780	Adhesive in shear
0·4	1 690	229 000	6 850	Adhesive in shear
0·4	1 830	248 000	7 440	Adhesive in shear
0·6	2 300	312 000	6 230	Steel in hoop tension
0·6	2 230	302 000	6 030	Steel in hoop tension
0·8	2 200	298 000	4 470	Steel in hoop tension
0·8	2 380	322 000	4 830	Steel in hoop tension

overlap longer than 1·5 in are very high. The corresponding tube wall thickness at this pressure is approximately 0·1 in. These tests have therefore been limited to bond lengths of less than 2 in and the results (Table 9) show that using the Araldite X33/1202 adhesive very high average shear stresses at failure can be achieved, even though the glue line thickness varies between 0·1 and 0·045 in.

The results of a series of tests on vessels stored under ISAT series B conditions are shown in Table 10 from which it will be noted that there is no deterioration after 15 cycles, whereas the lap joints previously described began to deteriorate immediately.

TABLE 9

Hydrostatic Pressure Test to Failure of End Fitment Bonds (6·5 in Diameter Pressure Vessels)

Failing pressure (lb/in^2)	Bond overlap length (in)	Adhesive shear stress at failure (lb/in^2)
1 750	0·5	5 850
1 800	0·5	6 050
4 600	1·0	7 480
4 750	1·0	7 730
4 900	1·0	7 930
4 500	1·0	7 310
4 600	1·0	7 510
6 000	1·5	6 700
5 850	1·5	6 350
7 600	2·0	6 360

Evaluation of full size pressure vessels (*11 to 24 in diameter*)

Results of further tests to assess the failing stresses of the end ring adhesive (Araldite X33/1202) using several combinations of diameter and end bond length are given in Table 11, where it is seen that the failing stresses achieved with large diameter vessels show little change from those obtained in small scale tests.

In certain circumstances it is necessary for a pressure vessel to withstand pressure and bending loads at elevated temperatures, and the results of an investigation of this problem are presented in Fig. 33, where it is seen that certain adhesives performed satisfactorily at much higher temperatures than others. This work, which was carried out in 1961, shows that the Bloomingdale HT 424

TABLE 10

Storage Tests of 6·5 in Diameter Pressure Vessels at ISAT 'B' Conditions

Period of storage (cycles)	Failing pressure (lb/in²)	Steel wall stress (tons/in²)	End attachment adhesive stress (lb/in²)
Nil	3 600	71·6	5 860
	3 800	75·5	6 180
	3 400	67·7	5 530
8	3 600	71·6	5 840
	3 700	73·5	6 020
	3 550	70·6	5 760
13	4 000	83·0	6 500
	3 600	71·6	5 800
	3 350	66·6	5 450
15	3 150	62·6	5 110
	3 700	73·5	6 020
	3 650	72·5	5 930

TABLE 11

Additional Hydrostatic Tests of End Fitment Bonds (Greater than 6·5 in Diameter)

Tube diameter (in)	Bond length (in)	Failing pressure (lb/in²)	Adhesive shear stress at failure (lb/in²)
9·5	2·0	4 200	5 000
11·75	2·0	4 400	6 500
11·75	1·875	3 900	6 110
11·75	1·375	3 550	7 580
11·75	1·25	3 000	7 050
11·75	1·875	4 100	6 420
23·5	3·0	2 800	5 420
23·5	4·0	3 600	5 220
23·5	4·0	3 450	5 070
23·5	4·0	3 500	5 100

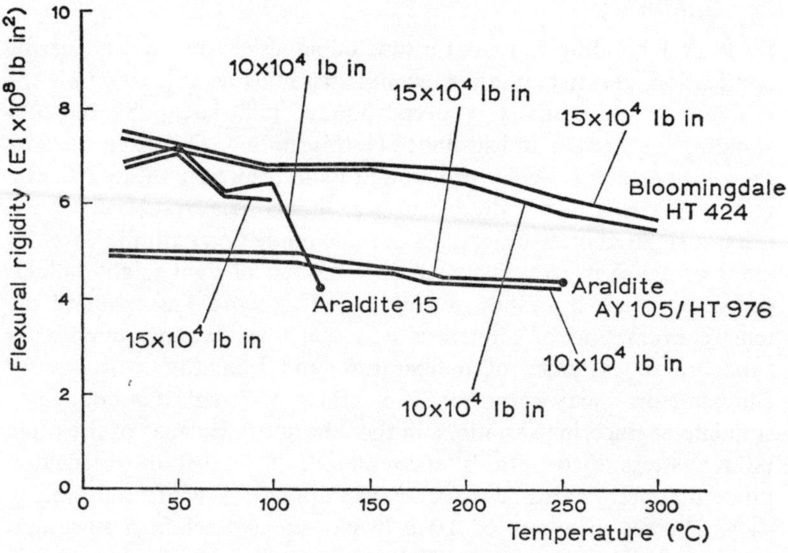

Fig. 33. Structural tests on tubes using different adhesives

adhesive performed satisfactorily up to temperatures of 300°C, and there is little doubt that this figure could be raised substantially with the use of more recently developed materials. Finally, Plate 13 demonstrates that correctly designed strip laminate vessels, when pressurised to destruction, behave as homogeneous metal bodies, giving a conventional hoop burst at 100 per cent of the uniaxial tensile stress for the metal used in their construction.

Conclusion

Destructive methods of assessing adhesive bonds have played an important part in the development of the IMI strip laminate process for the manufacture of tubular pressure vessels. Laboratory tests have been used to obtain comparative information on the performance of different adhesive formulations, but must be related to scale tests on typical structures before their results can be used as design data.

It is also seen that there can be no hard and fast rule for the evaluation of adhesives for any particular application, and it is hoped that the experience gained in the development of this process may prove useful to other investigators.

Summary

Structural bonding implies the use of adhesives in an engineering application, and the adhesive technologist must be able to satisfy the design engineer that the proposed bonded joint is capable of withstanding the stresses and strains to be transmitted. This paper reviews the methods available for providing this information with particular reference to single and double overlap joints, peel tests and torsion tests; it is based on work carried out in conjunction with the development of a novel process for the production of lightweight tubular pressure vessels by adhesive bonding. This work has required extensive evaluation of numerous adhesive formulations, often after exposure to extremes of temperature and humidity, with special emphasis on epoxy resin based materials. Although it is difficult to simulate engineering situations in the laboratory because of the often unique stress distributions encountered, it is demonstrated that modern adhesives can be successfully employed to withstand average shear stresses in excess of 4 000 lb per square inch in a structural application. It is also shown that correct selection of adhesive will enable a bonded structure to withstand considerable loads at temperatures in excess of 200°C.

Appendix A

The I.S.A.T. (Intensified Standard Alternating Trials) cycle tests have been laid down in the United Kingdom for uniformly testing equipment scheduled for use by the Armed Forces. The Series B cycle is the most severe, and one complete cycle is claimed to be equivalent to 9 weeks continual tropical storage. Details of this cycle are given below:

I.S.A.T. Series B Cycle

115°F	95% R.H.	2 days
140°F	60% R.H.	1 day
Cooling to ambient		1 day
167°F	No added moisture	8 hours
Cooling to ambient		16 hours
115°F	95% R.H.	1 day
Cooling to ambient		1 day

The complete cycle lasts one week and then is repeated automatically.

References

1. *The Strength of a Lap Joint*, Greenwood, L. (See p. 40).

Some Methods of Non-Destructive Testing of Adhesive Joints

W. THOMPSON, C.Eng., C.G.I.A., A.F.R.Ae.S.
FORMERLY A.I.D. LABORATORIES, HAREFIELD

Thermal test methods

A programme of work which is being undertaken at A.I.D. Laboratories has as its primary objective, the development of a reliable non-destructive method of assessing the quality of adhesive bonded joints of all types. The results of our work to date indicate that areas of non-adhesion, thick and porous adhesive layers, and in certain instances, areas of very weak adhesion can be detected relatively easily. The basic problem of differentiating between weak and strong joints, however, remains unsolved. In the case of metal to metal bonded joints, for example, a Redux bonded shear test specimen prepared strictly in accordance with the requirements of Specification D.T.D. 775 will normally have a failing load in excess of 1 800 lb; if however the pickling treatment of the adherends is omitted the failing load may easily fall below 500 lb, and as yet there is no non-destructive testing equipment which will detect any difference.

Of the various testing devices developed at Harefield, one of the first is illustrated in Fig. 34. This consists essentially of an extensometer, having two vertical arms one of which is rigidly attached to and the other pivoted to a horizontal member. The two arms terminate in knife edges which rest upon the specimen under test, so that any expansion or contraction of the specimen causes the pivoted arm to move. This movement is measured by a sensitive electronic comparator capable of measuring movements of 0·00001 in. On the underside of the horizontal member is fitted a mirror, and below this a row of low voltage electric lamps which provide sufficient heat to raise the temperature of the acting length of the specimen through a few degrees Celsius. The current fed to the lamps may be varied by means of a transformer, and during the test the specimen rests on a water cooled platen. The rise in temperature of the upper surface of the specimen is measured by means of a thermocouple.

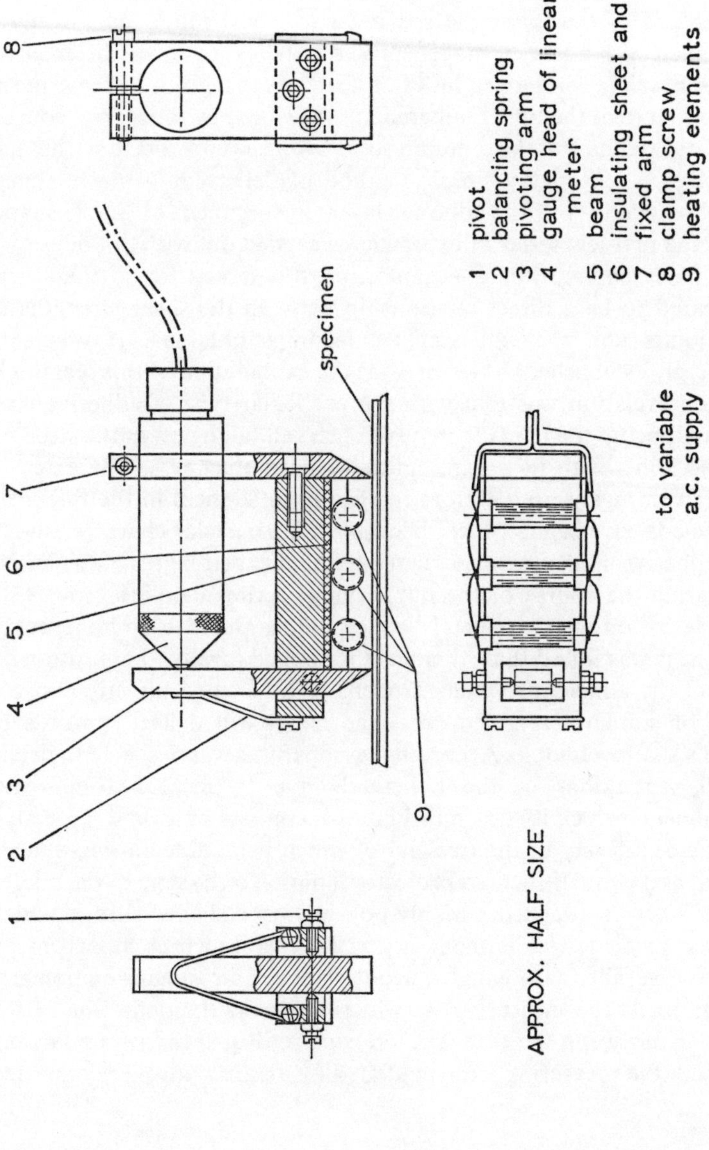

Fig. 34. A.I.D. bond tester (2 in model)

This device provides a useful means of detecting thick and porous adhesive layers, because for a given input of heat the temperature gradient between the upper and lower adherends will depend largely upon the rate at which the heat passes through the adhesive. In the case of a thick or porous adhesive layer the temperature gradient will be relatively large, and due to the unequal expansions of the two adherends of the joint, the specimen will bow more than would a specimen with a thin homogeneous adhesive layer. Since the extensometer reading, for a given input of heat, or for a given increase in the temperature of the upper adherend as measured by the thermocouple, will depend upon the amount of bowing which occurs, the test provides a reasonably simple method of determining the thickness and homogeneity of the adhesive layers in specimens of simple shape.

In the first few series of experiments carried out with this device on specimens having different joint strengths, it was found that there appeared to be a direct relationship between the shear strengths of the joints and the extensometer readings obtained. It was subsequently established, however, that the explanation of this seemingly good correlation was that with the particular type of adhesive used for making up the joints there was a marked falling off in the strength of the joints with increasing adhesive layer thickness. The extensometer readings were therefore indicating differences in the thickness of the adhesive layers which, because of a particular characteristic of the adhesive used, were also related to the shear strengths of the joints.

During the course of the numerous experiments carried out with this device on metal to metal bonded joints, and also on honeycomb panels, it was found that when heat is applied locally to a position on one of the adherends which happens to be located directly above a void or similar unbonded area, the metal will deflect upwards to form a shallow bubble. Accordingly, apparatus was designed to detect small separations of the adherends due to local heating. This equipment[1] which is shown in Plate 14, consists of a rigid laminated plastic centilever, to the free end of which is mounted a micrometer screw, and two self-focused projector lamps. To the stem of the micrometer screw is attached a highly polished cone-shaped shroud which houses an air jet made from Invar. This jet which is connected by means of a tube to the gauging head of a Solex air gauging equipment, functions as the measuring jet which monitors the deflection of the workpiece during the test. The pressure and quantity of air flowing through the measuring jet is regulated by an air controller connected to a small compressor. The air controller incorporates a water manometer and a fall of 14 in in the level of the water in the manometer tube represents a decrease in the distance between the measuring jet and the specimen under test of approximately 0·005 in.

The self-focused projector lamps are arranged to irradiate a roughly circular area approximately one inch in diameter surrounding the point on the specimen at which the stream of air from the jet is impinging. The intensity of the illumination can be varied by varying the filament voltage.

In order to carry out a test the specimen is positioned under the air jet, and the distance between the jet and the specimen adjusted by means of the micrometer screw until the level of the water in the manometer tube reaches the zero position. Ideally the surface of the specimen should be flat and normal to the direction of the air stream but in practice it was found that the equipment functioned quite satisfactorily on curved surfaces. The lamps are then switched on, and if there is a void or unbonded region the adherends immediately commence to separate. The upward deflection of the adherend being heated diminishes the gap between it and the air jet, and the level of the water in the manometer tube begins to fall. If, however, the adherends are bonded together no separation occurs, and the level of the water in the manometer tube remains stationary.

This apparatus proved to be quite effective for testing finished parts having painted or anodised surfaces. At the outset it was considered likely that in the case of specimens painted in different colours it would be necessary to vary the current being supplied to the lamps to accommodate the different absorptivities. This was subsequently found to be unnecessary however: presumably the organic binder in the different paints absorbed a high proportion of the infrared radiation in the beams. The transformer setting, which for a matt black painted surface gave an increase in temperature of approximately 35°C in the irradiated zone of the specimen, was found to be quite satisfactory for other colours including gloss white. Moreover, the inspection of finished components presented no difficulties since the avoidance of any contact between the component and the testing equipment eliminated the possibility of scratches and similar mechanical damage.

The performance of the equipment was less satisfactory, however, in the case of metal to metal joints and honeycomb panels having brightly polished skins of 20 S.W.G. and thicker. The high reflectivity of the polished surfaces coupled with the relatively high rate of heat dissipation resulted in temperature rises in the irradiated regions of less than 15°C, and these were not sufficiently large to cause separation of the joints. In order to overcome this difficulty the apparatus illustrated in Plate 15 was constructed. With this device the projector lamps are replaced by a high frequency induction heating unit. The air jet and shroud are fabricated from a proprietary ceramic material and are attached to the end of a micrometer screw which in turn is mounted upon a ceramic tripod. Secured to the legs of the tripod is the water cooled pancake coil of a 1 kW induction heater.

One interesting feature of this type of test is that the severity of the test is automatically increased in the case of areas of non-adhesion and weakly bonded areas, because the heat cannot be dissipated so readily through the adhesive layer into the lower adherend, or in the case of honeycomb panels into the cores. Temperature measurements made at the surface of a honeycomb panel containing small areas of non-adhesion, due to the inclusion of pieces of polythene sheet material between the skins and the adhesive, revealed that in the correctly bonded regions, the temperature rise for a certain instrument setting was 18°C in 10 sec and 30°C in 15 sec. The corresponding values over the unbonded regions were 20°C in 10 sec, and 46°C in 15 sec. The temperature differences were not so marked in the case of strong and weakly bonded areas but in several instances when the induction heating method was being employed, failure of a very weakly bonded joint occurred during the test.

This same effect was made use of in another testing equipment[2] which was developed primarily for the purpose of inspecting bonded joints between resin impregnated fibreglass sheet materials and aluminium alloy plates. In this particular project the components consisted of aluminium alloy plates approximately 12 in × 30 in × 0·5 in thick to both sides of which had been bonded 0·060 in thick sheets of fibreglass. Since there was no suitable destructive test apparatus available for determining the strength of the bonded joints some simple tests were carried out in which the cutting edge of a thin wood chisel was driven between the fibreglass sheets and the metal plates. With some of the components tested in this manner it was found that the fibreglass could be detached quite readily from the metal, and after a small area of the bonded joint had been broken, the fibreglass sheet could be peeled away from the metal by hand. With other components tested in this manner, however, it was found

PLATE 14 (*Crown copyright reserved*) See page 78

PLATE 15 (*Crown copyright reserved*) See page 80

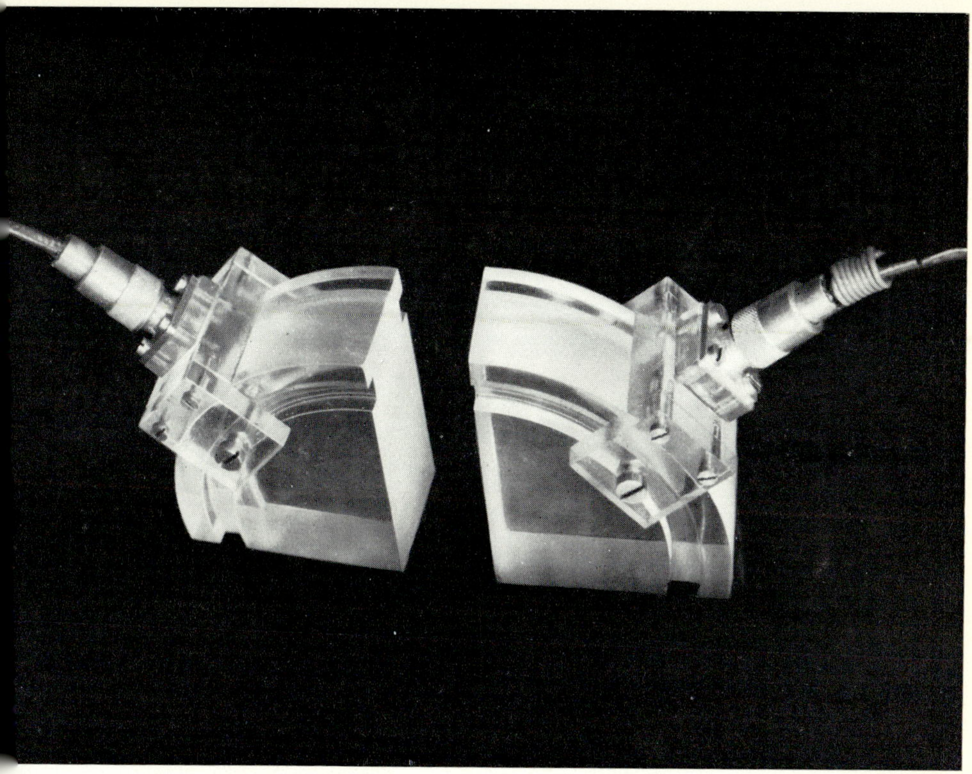

PLATE 16 (*Crown coypright reserved*) See page 84

PLATE 17 (*Crown copyright reserved*) See page 84

that the fibreglass could not be separated from the metal and the use of additional force merely resulted in local ruptures and exfoliation of the fibreglass sheets.

Components from both groups were then tested with the projector lamp and air gauging equipment but even after prolonged exposures to the heat both weak and strong joints appeared to be intact. It was noted, however, that if the test was carried out in a region where the joint between the fibreglass and the metal had been broken, there was a very rapid rise in temperature and within a few seconds scorching of the resin impregnated fibreglass sheet occurred. Further tests were therefore carried out on both weak and strong joints, in which the temperatures of the fibreglass sheets were measured in the regions being heated. The results of these tests showed that the heat arriving at the irradiated zone of the fibreglass sheet passed through the adhesive and into the metal beneath it, a little more rapidly in the case of the strong joints. Accordingly experi-

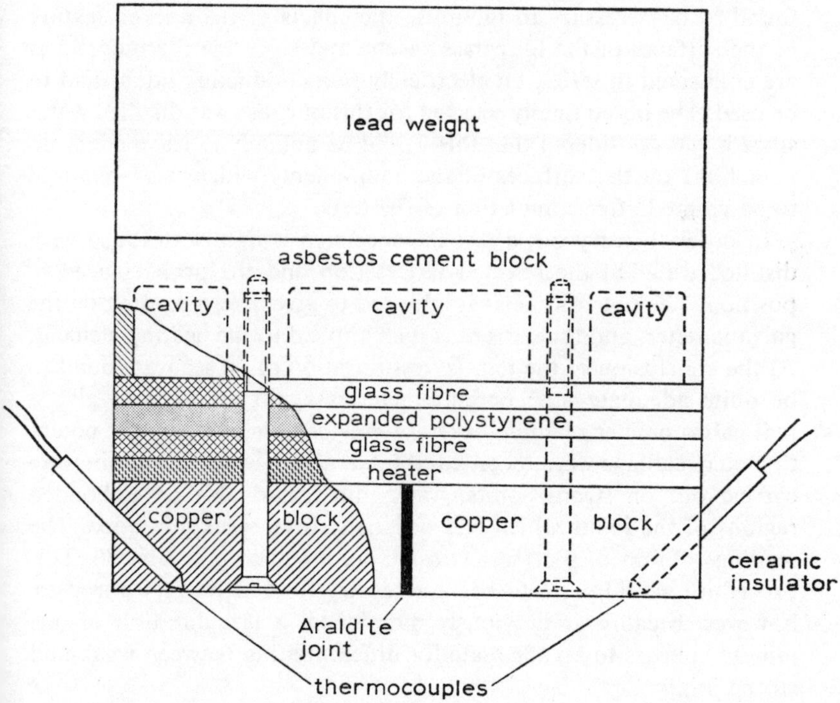

Fig. 35

ments were undertaken with various types of thermal probe and one arrangement which was found to function satisfactorily is illustrated in Fig. 35. This consists of two copper plates 1 in × 0·75 in × 0·018 in thick, adhesively bonded together along one edge, and each containing a thermocouple embedded in its lower surface. Over the upper surfaces of these plates is a low voltage heating element encased in sheets of mica, and above the heating element a layer of expanded polystyrene which in turn is encased in layers of mica or fibreglass. This assembly is screwed to an asbestos cement block which is recessed to reduce the surface area, and thereby minimise the upward flow of heat. Attached to the upper surface of the asbestos cement block is a lead weight weighing about 2 lb. The thermocouples are connected in series and the cold junctions immersed in a container of melting ice.

A normal potentiometric method is used to measure the temperatures of the copper plates in the probe assembly during the test. The use of a liquid couplant between the probe and the workpiece was found to be necessary to minimise the effects of the uneven texture of the surfaces of the fibreglass sheets, and since the thermocouples are connected in series, an electrically non-conducting liquid had to be used. The liquid finally selected for this purpose was distilled water since it was considered that this would be unlikely to leave contaminant films on the surfaces of the components which were required to be painted after completion of the tests.

In order to carry out a test the specimen is lightly swabbed with distilled water at the selected test station and the probe placed in position. The potentiometer is adjusted to give a zero reading on the galvanometer, and the current is then applied to the heating element. At the conclusion of the test—a test duration of 60 sec was found to be quite adequate—the potentiometer is again adjusted to give a null galvanometer reading. Several series of tests in which the potentiometer readings were recorded at intervals of 15 sec for 3 min were carried out on strong joints, weak joints, and also on unbonded regions of the joints which had originally been weakly bonded. The readings obtained, plotted against time, are shown in Fig. 36. This procedure would seem to be unnecessary for routine test purposes, however, because as previously mentioned, a test duration of one minute appears to be adequate for differentiating between weak and strong joints.

On several occasions since, the equipment has been put into service

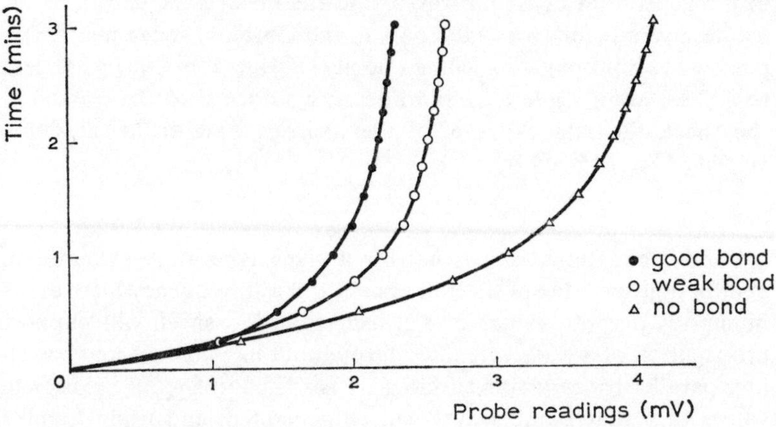

Fig. 36

and components whose bonded joints appeared to be weak, have been detected, and in every instance the non-destructive test results have later been confirmed by the chisel test method. As yet, however, no conclusive proof has been obtained that differences in joint strength are actually being detected, and it is possible that the higher instrument readings which are obtained on weak joints may result from some abnormality, such as porosity in the adhesive layer, which in turn results in a weak joint.

Ultrasonic test methods

Turning now to ultrasonic test methods, it has been our experience that the limitations of the thermal test methods, apply also to the various ultrasonic bond test methods in common use. Defects such as voids, unbonded regions, and thick adhesive layers can be detected quite easily, and indeed frequently more speedily than is possible with the thermal methods. But joints which are weak because of weak adhesion between the adhesive and the adherends cannot be detected.

Our programme of work at Harefield has been mainly concerned with investigating the usefulness, in the inspection of bonded joints, of various designs of ultrasonic probes. In the case of thin metal to metal joints, and honeycomb panels, the thicknesses of the adherends usually approximate to the wavelength of the ultrasound. This results in the generation of Lamb waves, and the metal adherends and adhesive layers themselves, take part in the vibratory movements. If a plane wave is propagating with a velocity (V_I) and is obliquely incident to a sheet at an angle θ from a line drawn normal to the surface of the sheet, the rate of travel of the incident wave front, or phase velocity (Vp) is given by:

$$V_p = \frac{V_I}{\sin \theta}$$

Lamb's equations[3] show that for a given type of sheet material, and for a given value of sheet thickness and test frequency there exists numerous discrete values of V_p such that the sheet will support propagation of a wave extending throughout its thickness and travelling parallel to its major surfaces. It is, of course, only at discrete values of θ that Lamb waves will be generated, and using Lamb's original equations, Firestone and Ling[4] calculated the phase velocities of the first few symmetrical and asymmetrical modes in aluminium sheets. The diagrams they published, showing phase velocity plotted against frequency times sheet thickness, are useful for deriving the appropriate angle of incidence for a particular mode in plain sheets. In the case of bonded joints made up from materials of widely differing acoustic impedance, however, it has been our experience that the suitable angles of entry and exit of the ultrasound are best determined experimentally. For this purpose, a pair of variable angle probes of the type illustrated in Plate 16 are employed. Each probe consists of a perspex shoe in the form of a quadrant, along the curved surface of which runs a suitably shaped carriage which is spring-loaded to hold it firmly in contact with the shoe. The crystal is mounted on the upper surface of the carriage, and thin films of oil between the crystal and carriage, and the carriage and the quadrant shoe ensure acoustic coupling between the crystal and the shoe.

In use, the two probes are mounted on one of the specimens to be tested, with one crystal functioning as the transmitter, and the other as the receiver. The carriages are slowly moved around the quadrants until the angles of incidence and exit are found which give a reasonably strong signal. Since the crystals are easily changed it is a simple matter to replace them with others of different type and frequency if this is found to be necessary. The angles are then measured, and since the quadrant probes would be too cumbersome to use in routine testing, a probe assembly of the type illustrated in Plate 17 is constructed. This consists of a suitably shaped perspex block with the crystals housed in recesses the bases of which are inclined at the angles determined in the previous experiment.

With this type of probe assembly it is found that whenever it is passed over an unbonded region there is a massive increase in the amplitude of the received signal. This method has the advantage of not being particularly time consuming, since quite large components can be scanned in a relatively short space of time. One difficulty which has been encountered, however, stems from the fact that minor differences in the thicknesses of the adherends being scanned sometimes result in very considerable variations in the amplitudes of the signals received. It appears to be possible to minimise these effects by varying the distance between the transmitter and receiver crystals and this particular feature of the test is being investigated at the present time.

References

1. A.I.D. Report N.D.T./2007.
2. A.I.D. Report N.D.T./56/65.
3. *Proc. Roy. Soc.*, Series A Volume XCIII, p. 114.
4. US Patent No. 2,536, 128, Jan., 2, 1951.

Adhesion in Electronics

A. W. HENDERSON, B.Sc., Ph.D.
THE PLESSEY COMPANY LIMITED

Electronic components depend on a science of materials whose adjoining interfaces are subject to stress by electric potentials. Because reliability in component performance is related to the strength of the bonds between the dielectrics and conductors involved, a case is at once apparent for introducing adhesion to those interested in electronics; and since semiconductor physics has familiarised engineers with electronic theories of valency, there is some probability that they will make contributions to a precise science of bonding and adhesion.

This paper deals, by way of introduction, with some developments in typical electronic devices, with emphasis on bonding and its effects. Consideration of semiconductors then leads to the introduction of several propositions on the present electronic theory of adhesion.

The electronic components which were made without radical change between the wars and can now almost be called traditional, were resistors, condensers, inductances and vacuum tubes or valves. But though basic conceptions did not change much, advances in technique of manufacture were continually being made and over the years, the sizes of individual devices have been much reduced.

An audience trained mainly in other disciplines is entitled to ask why there should be reduction of size since it obviously results in closer tolerances and intricacy in manufacture. But, electronic instruments carry out increasingly complicated functions and, for example, a three-valve radio receiver of the thirties has been replaced by the three colour television set of today. In the same domestically acceptable volume will now be packed possibly one thousand times the number of components and the assembly process has required fabrication of as many joints between them; therefore the components must be small. Those who have actually made a soldered joint will realise the difficulties involved in ensuring reliable connections on this scale. There has been an incentive therefore to miniaturise and combine the functions of many components into one module, and by carrying out assembly automatically, to eliminate human variability. It is of interest to note some aspects of the development of conventional components and especially condensers, where a knowledge of bonding has played a significant part in the advances made.

A condenser, you may recall, can be represented diagrammatically by

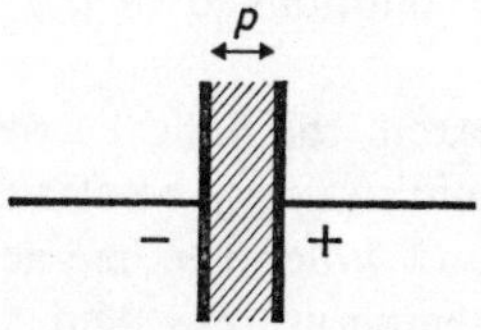

Fig. 37. Diagram of a condenser

two flat conducting panels separated by a distance *d*, by the dielectric (Fig. 37). Size reduction without decrease of capacity was difficult with conventional materials, say metal foil and waxed paper dielectric. It can be shown that the capacity of a condenser $C \propto \frac{1}{d}$; therefore to reduce bulk, reduction in thickness of dielectric would seem to be indicated. But exceptionally thin paper soon ceases to be self-supporting and besides can be ruptured by electric stresses. Another approach was required.

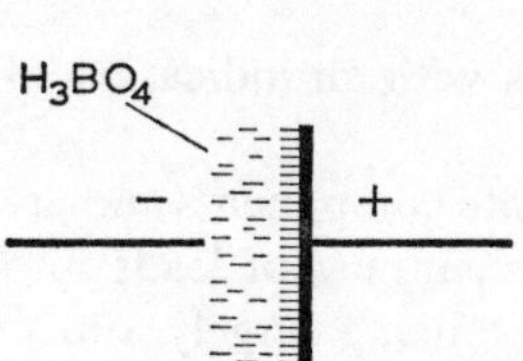

Fig. 38. Diagram of an electrolytic condenser

Anodic treatment of aluminium in the presence of a weak electrolyte such as boric acid, creates a tenaciously adhering extremely thin oxide layer on the metal. This remains intact especially in the presence of an applied potential and its thickness is approximately 13·5 Å/volt.

Using the electrolyte as negative terminal and aluminium metal as the opposing positive plate, it has been possible to produce large capacity condensers of small size because of the extremely thin dielectric layer (Fig. 38).

Such condensers must obviously be entirely sealed against escape of electrolyte but early samples occasionally exploded in use. Investigations showed that chloride, present as an impurity in the electrolyte, would appear at the anode and attack the oxide layer next to the aluminium. Rupture of the bond and concentrated electron flow at that point then led to excessive local heating and disaster. Close control of chloride in electrolyte is now maintained in the manufacturing process.

It is interesting to speculate, to what extent the bond between the natural or thickened oxide layer on aluminium may be attacked by chlorides, always present as impurities, and which become active in *adhesively* bonded joints at elevated temperatures. And though degradative attack on the inorganic component of a joint has not apparently been reported, it is known that ionic impurities catalyse the thermal degradation of organic polymers.[1, 2]

Plastic and paper films would conventionally be used in metal foil condensers down to a dielectric thickness of perhaps 0·00025 in. The aluminium oxide electrolytic barrier layer would be approximately 100 Å. It has lately been possible also to use tantalum metal with an electrolyte of sulphuric acid further to reduce size. The excellent bond between tantalum oxide and the base metal is interrupted only by surface impurities so that absolute cleanliness is necessary in production.

Plate 18 shows condensers of various types with an indication of size.

The devices shown, except *d* and *e*, are discrete components and are connected by soldering, welding, brazing or crimping of leads into practical equipment, wire linking conductors being largely superseded by printed wiring connections.

Plate 19 shows a modern printed wiring panel with components mounted, and to produce it, a formidable array of bonding problems has had to be overcome, e.g. cohesion of the paper or glass-fibre based laminate, adhesion of conductor metal to laminate, adhesion of solder or electroplating to conductor often with adverse environmental exposure.

Whilst printed wiring will continue to be used for many years, its disadvantages become noticeable as greater reliability is sought. The insulating carrier, typically a resin-bonded board, absorbs moisture quite easily and the ensuing dimensional change tends to shrug off metal bonded to its surface. This effect is exaggerated by differences in thermal expansion between plastic materials with a linear coefficient ranging from 50 to 120 $\times$ 10^{-6}/°C and metals at 10 to 30 $\times$ 10^{-6}/°C. Disturbance of electrical performance inevitably follows mechanical stress and strain. Glass and ceramics however, offer ready-developed, stable base insulants, and small samples are remarkably robust.

Borate alkali-free glass has a particularly low expansion coefficient and ceramics combine toughness and good heat-conductivity.

Table 12 compares relevant variables:

TABLE 12

Comparative Properties of Materials of Construction for Electronic Circuitry

Property	S.R.B.P.	Epoxy glass laminate	Electrical glass
Water absorption (%w/w on $\frac{1}{8}$ in thick sheet 24 hours)	0·2—4·5	0·05—0·30	0
Impact strength (ft lb/in)	4—8	10—25	0·1—0·3
Thermal conductivity (10^{-4} cal/sec cm² °C)	1·8	7	1200
Thermal expansion (10^{-5}/°C)	1·4—3·0	1·0—1·2	0·4—0·9
Volume resistivity (Ω cm)	10^{10}—10^{13}	10^{15}	$10^{13\cdot5}$

Property	Alumina	Aluminium	Copper
Water absorbtion (%w/w on $\frac{1}{8}$ in thick sheet 24 hours)	0	0	0
Impact strength (ft lb/in)	0·4	20	60
Thermal conductivity (10^{-4} cal/sec cm² °C)	2·4 $\times$ 10^4	5 700	2 900
Thermal expansion (10^{-5}/°C)	0·8	0·24	0·20
Volume resistivity (Ω cm)	10^{14}	2·5 $\times$ 10^{-6}	1·6 $\times$ 10^{-6}

An adhesive is actually found to be unnecessary between glass or ceramic base plates and metallic conductors, provided that they are produced in situ by vapour or chemical deposition. Introduction of organic material could only degrade such an arrangement.

It is of interest now to consider this bonding process as used in the manufacture of so-called 'Thin Film Circuitry'.

Rectangles about 1 in × 0·5 in in alkali-free, electrical grade glass 0·025 in thick are thoroughly cleaned in a variety of solvents ending with rinsing in deionised water and drying in refined air. Mechanical handling prevents contamination. The slides, clamped in carrier frames, are rotated under vacuum above a heated crucible of the material to be deposited, intervening masks ensuring deposition in the required pattern.

Conductors may be of gold or aluminium; resistances are built up with nickel/chrome vapour and insulants include glass, silica, etc. Table 13 indicates typical characteristics.

TABLE 13

Showing properties of t.f.c. resistors and capacitors

Resistors	
Material	Nickel chrome
Sheet resistance	200 Ω/square
Resistance range	10 Ω – 100 k Ω
Manufacturing tolerance	±0·1% to ±10%
Temperature coefficient	±25 p.p.m./°C
Coefficient matching	5 p.p.m./°C
Stability	0·1%/1 000 h at 70°C full load
Operation frequency	1 000 MHz approx.
Dissipation	1·5 W/in² of substrate
Capacitors	
Material	Aluminium-silicon monoxide
Capacitance	60 000 pF/in²
Range of capacitance	10–10 000 pF
Manufacturing tolerance	±1% to ±20%
Temperature coefficient	±100 p.p.m./°C
Stability	0·1% 1 000 h at 70°C
Power factor	0·001 at 100 kHz

Under the conditions fractionation of alloy constituents may occur and it is particularly noticeable that nickel has a different evaporation rate than chromium. But with deposition of successive small samples of a specially prepared alloy, nickel-chrome with an excellent thermal resistance coefficient, can be obtained.

This metal automatically adheres well so long as the receiving glass temperature is about 200°C and the pressure 10^{-4} torr. Now clearly high temperature and low pressure ensure the removal of moisture and other volatile contaminants from the glass surface, but surprisingly further reduction of pressure actually reduces the effectiveness of the metal/glass bond. It is now thought probable that some chromium atoms may be oxidised during flight, and that the chromic oxide so formed on impinging on silicious material may react with the possibility of covalent bonding (see Fig. 39). Shortage of water may interfere with this process.

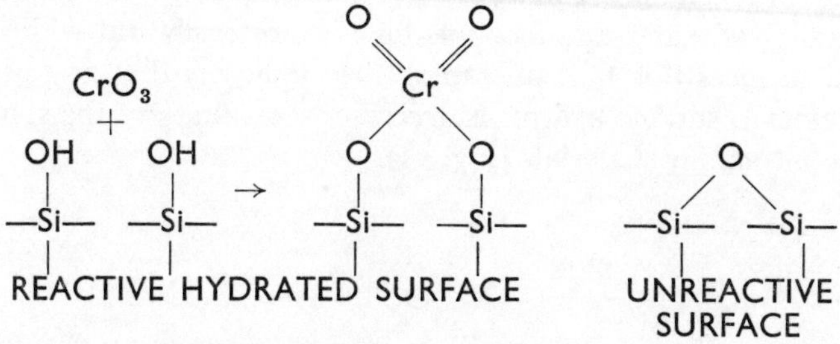

Fig. 39. Possible method of reaction of chromium on a silicious surface

With the possibility of covalent bonding between glass and chromium and metallic bonds in the nickel/chromium alloy, good joints can be expected (Fig. 39).

In order to test the bond between glass and deposited metal, wires are soldered on and pulled off normal to the surface. Strengths of at least 1 000 lb/in^2 mean that the glass generally breaks before the bond. Good adhesion requires compatibility of succeeding layers. Molten solder will not adhere to nickel/chrome but if nickel is deposited on the nickel chrome and a light flash of gold on the nickel, soldering with selected tin alloys is possible (Fig. 40). For the same reasons deposition of gold conductors on glass requires a priming layer of nickel/chrome.

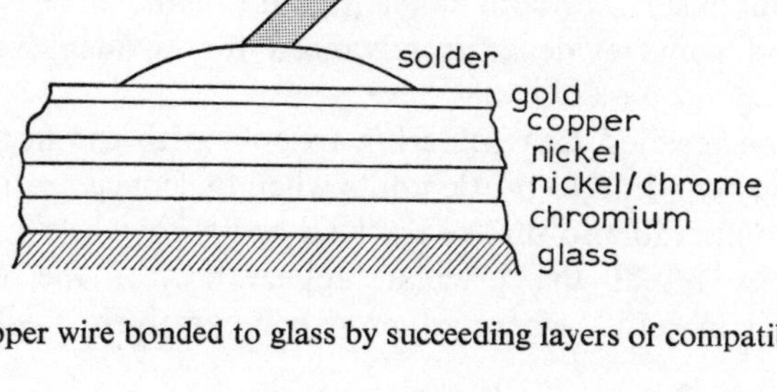

Fig. 40. Copper wire bonded to glass by succeeding layers of compatible material

It is of interest to compare these practical results with the work of Weaver[3] who showed that growth of iron oxide preceded improved adhesion of the evaporated metal on glass; aluminium reacts similarly. As might be expected on this hypothesis, gold always adheres badly. These results are in keeping with the view that covalent bonding has occurred to the glass.[4]

Evidence of chemical reaction by organic compounds on silicious surfaces is available. Hockney[5] shows that with fully hydrated silicas, adsorption of water and alcohols takes place easily but with dehydrated surfaces this falls off rapidly. He indicates that any process that removes surface hydroxyls *decreases* the affinity of the solid for electron donating materials (Fig. 41).

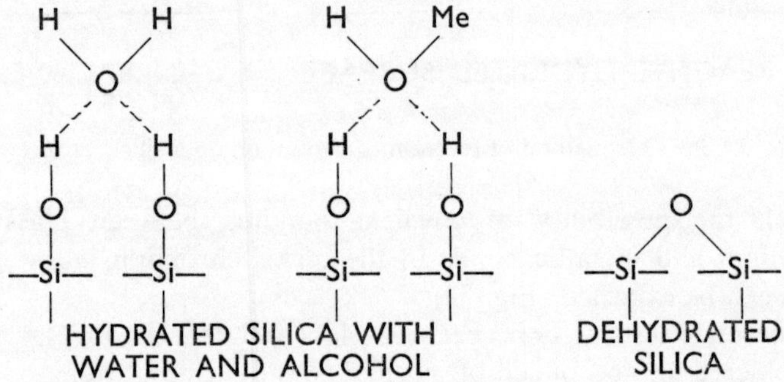

Fig. 41. The reactivity of hydrated silica with water and alcohol

High temperatures however, are not necessary for a purely chemical reaction to take place. On alumina, at room temperature, carbon disulphide is decomposed to its elements.[6] By the application of infrared spectroscopy methods for the study of adhesive/glass interaction Krotova and Morozova[7] have shown that reaction between substrate and adhesive may continue for some time.

This evidence tends to show that opportunities for chemical bonding on suitable surfaces with conventional organic adhesives are not lacking and some evidence is advanced for covalent bonds with chromium metal on a silicious base.

It may be recalled that soldering to gold with the normal 40/60 lead/tin alloy can lead to brittle joints when the concentration of gold becomes comparable to that of the lead or tin anywhere in the joint. Micrographs reveal the acicular appearance of the compound (Au/Sn:1:1) in joints where either a comparatively thick layer of gold was present, (e.g. 0·0002 in as used for switch contacts as opposed to 10^{-6} in necessary for corrosion prevention), or heating was insufficient to ensure gold dispersion. Some incompatibility in reported results,[8, 9] may be due to methods of test. It can be shown that with organic adhesives that a brittle material appears weak under peeling stress. The following peel test figures obtained recently tend to confirm this impression and emphasise differences due to undispersed gold, whereas lap or direct tension joints are unlikely to discover the weakness.

With a thick layer of gold and a short heating cycle, brittle failure occurs (Table 14).

TABLE 14

The Strength of Gold-Solder Joints varying with Heating and Gold Thickness
(Peel strengths in pounds of 24 s.w.g. copper wire soldered to various substrates).

Substrate	*Temperature of soldering iron °C* 220°C	250°C
Copper	3·1	3·0
0·00005 in Gold on Copper	3·0	2·7
0·0002 in Gold on Copper	2·1 (Brittle failure)	2·2
Copper—Gold removed by wicking off with molten solder	3·2	3·1

Wire 24 s.w.g. hot tin dipped copper wire
Solder 18 s.w.g. 60/40 'Ersin Multicore' Solder
Soldering Iron 25 watt 'Solon' power supply adjusted to give different bit temperatures
Substrate 0·0015 in thick copper clad S.R.B.P.

In the assembly of electronic equipment where soldering to gold surfaces is necessary, the use of silver-containing solders ensures suppression of the unwanted brittle gold alloys. The reason for this may be quite simply due to better dispersion with higher temperature of application (360°C–380°C compared to 200°C–220°C) but this has not been investigated.

Thin film circuits are useful for building up interconnecting chains of precision resistors and capacitors and can carry terminals for lead-out wires or microcircuit elements.

Provided a reasonable vacuum is maintained and contamination e.g. from pump oils and excess oxygen is prevented, evaporated materials will automatically adhere well to the base and to each other.

The nature of the bond must range from covalent bonds in metal/metal alloys to mixed metal/oxide links and oxide/oxide structures with only secondary (van der Waals') forces and even interlocking structures contributing.

Stages in the design and layout of a thin-film circuit are as follows (Fig. 42).

A theoretical circuit diagram is drafted by a designer (a) and re-drawn in a suitable form (b). Connector pads in aluminium are decided on (c) and the complete layout shown at (d). A finished circuit is shown in Plate 20.

Now the area of a thin-film circuit is $1 \times 0{\cdot}5$ in and on to it can be evaporated 5–50 separate 'components'. But still smaller devices, microcircuits, each carrying out a large number of separate functions and only about 0·1 in square, may have to be bonded to the thin-film base. The highest packing density is achieved by the so-called 'flip chip' technique where aluminium/aluminium joints are made by a kind of projection welding using ultrasonic friction heating. Extreme precision is required for this operation—Fig. 43 and Plate 21.

It is worth noting that as many as 32 welded joints can be made at once by this method.

Apart from precise positioning, good bond quality is related to the purity and homogenity of the projections of weld metal. Pure aluminium readily permits friction welding to itself and is preferred as a conductor to e.g. copper, for its greater environmental stability.

I come now to microcircuitry where (by way of illustration) it is a practical and economic proposition to produce most of the functional components of a radio receiver on a 0·1 in square piece of silicon. Whilst many textbooks describe the process involved, it is appropriate to recall some of the elements of theory since it may have relevance to electronic theories of adhesion.

Silicon or germanium elements of Group IV and possessing four orbital electrons, display high resistance when grown as pure oriented single crystals. Minute (1 atom in 10^6) quantities of elements from Groups III or V may be added without disturbing the basic tetragonal structure, and lead in the first case to a shortage of electrons in the lattice and in the second to a slight excess. Neither structure has resultant charge but though the 'foreign' atoms remain

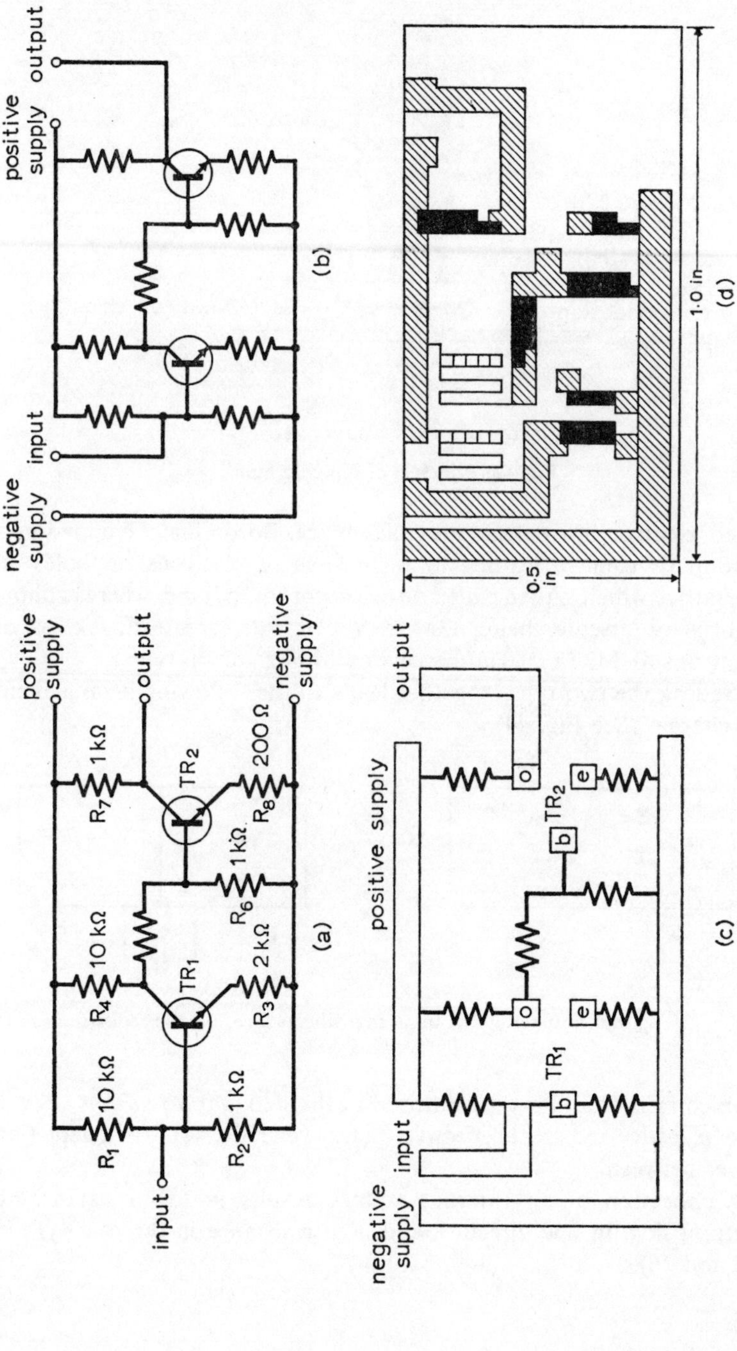

Fig. 42. Stages in the design and layout of a thin film circuit

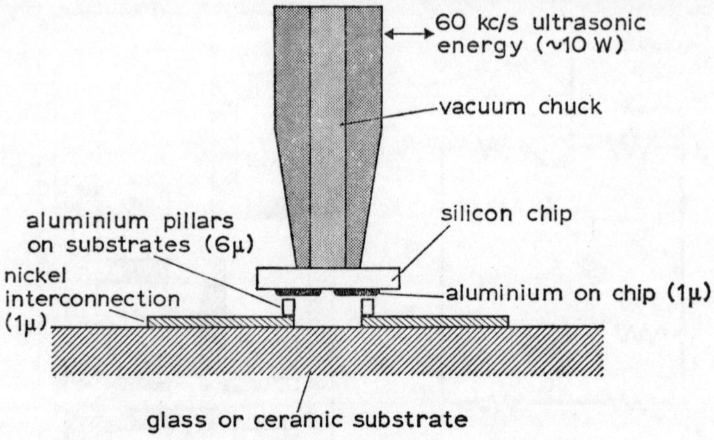

Fig. 43. Principle of flip-chip bonding

fixed, electron movements may take place. Boron may be added as a Group III element leading to a shortage of electrons or 'holes' in the lattice which is then called an 'acceptor' or 'p' type, whereas phosphorus or arsenic, being Group V elements, create an excess of electrons and the material becomes a 'donor' or 'n' type.

Setting the two types together leads to the following arrangement of charges (See Fig. 44).

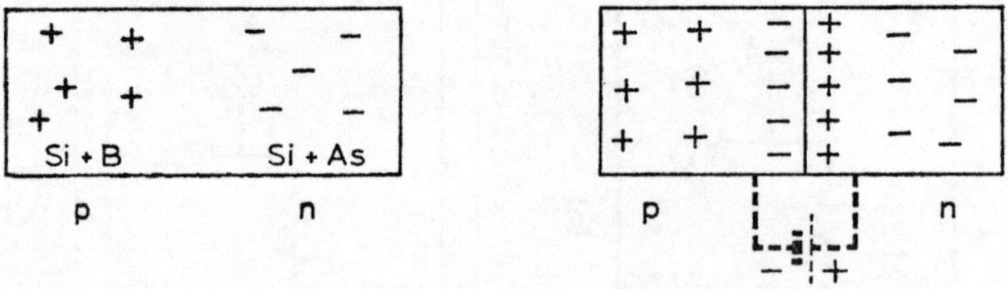

Fig. 44. (a) p and n material adjoining in a silicon slice; (b) the redistribution of charges that occurs

Electrons from 'n' type diffuse in a limited way to vacant sites in the 'p' lattice and create effectively a tiny cell (Fig. 44) with a depletion layer between.

Connection of an external voltage results in low resistance to current flow in one direction and high resistance in the other (Figs. 45 and 46).

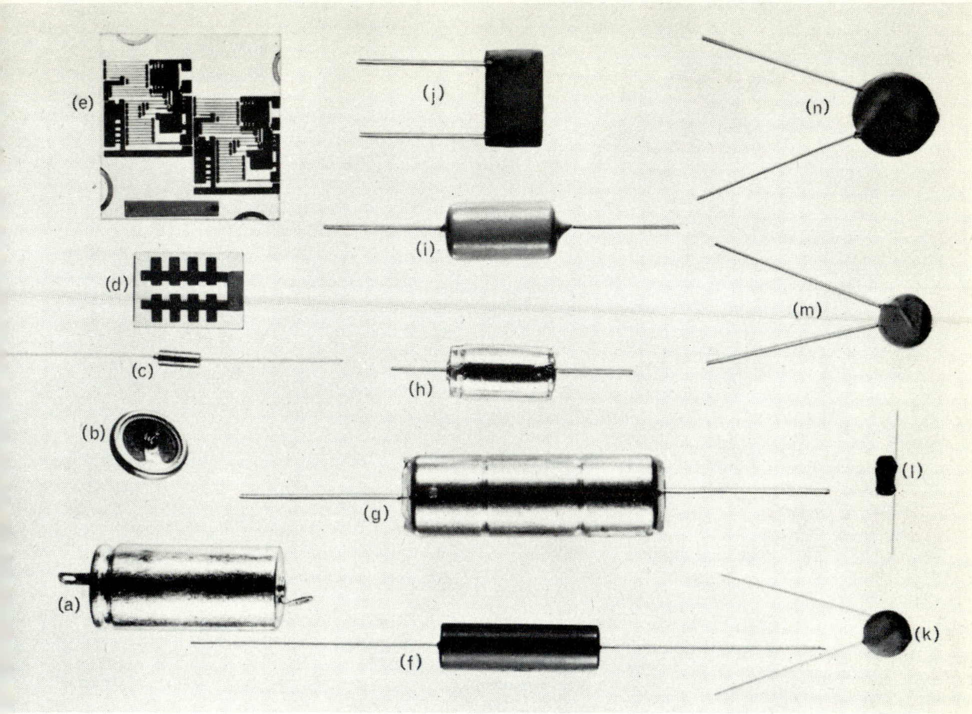

PLATE 18 Selection of modern capacitors. Scale: "b" is 1 in diameter

(a) Wet aluminium electrolytic, 50V 250μF
(b) Wet tantalum electrolytic, 70V 55μF
(c) Solid tantalum electrolytic, 35V 6.8μF
(d) Experimental thin film aluminium electrolytic, eight units each 10V 0·28μF
(e) Evaporated silicon monoxide capacitors with Ni/Cr resistors in active filter network, 20V 1500 to 2500pF
(f) Paper, 500V 0·1μF
(g) Metallised paper, 400V 2·0μF
(h) Polystyrene, 250V 0·050μF
(i) Polyester, 400V 0·33μF
(j) Silvered mica, 500V 800pF
(k) Low permittivity ceramic, 1500V 50pF
(l) Medium permittivity ceramic, 750V 330pF
(m) High permittivity ceramic, 3000V 5000pF
(n) Barrier layer (surface reoxidized) ceramic, 3V 2·0μF

(*Courtesy The Plessey Company Limited*)

PLATE 19 Mother board with provision for twelve similar size child boards (*Courtesy The Plessey Company Limited*)

PLATE 20 A typical thin film package (*Courtesy The Plessey Company Limited*)

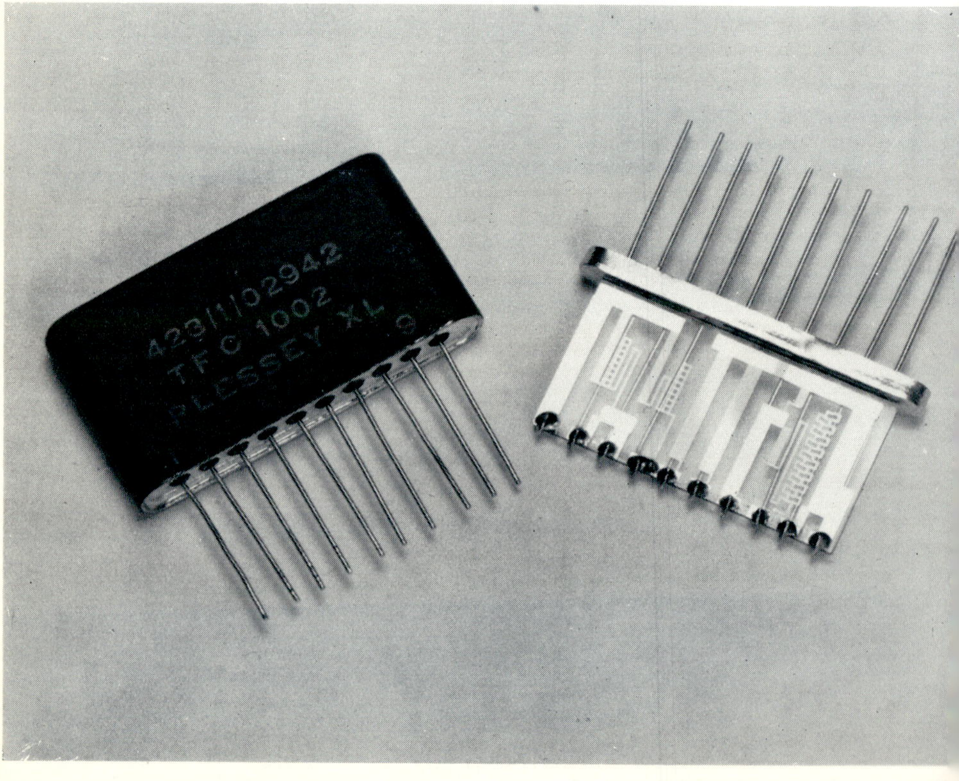

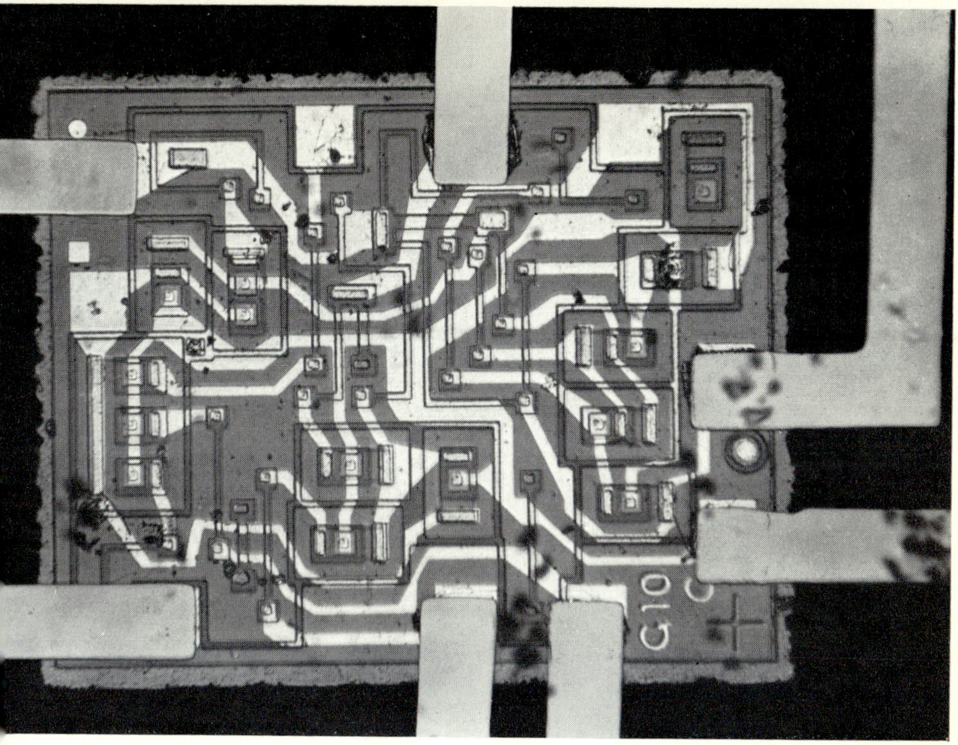

PLATE 21 Bonded chip seen through glass substrate (*Courtesy The Plessey Company Limited*)

PLATE 22 Circuit diagram and equivalent SIC chip (*Courtesy The Plessey Company Limited*)

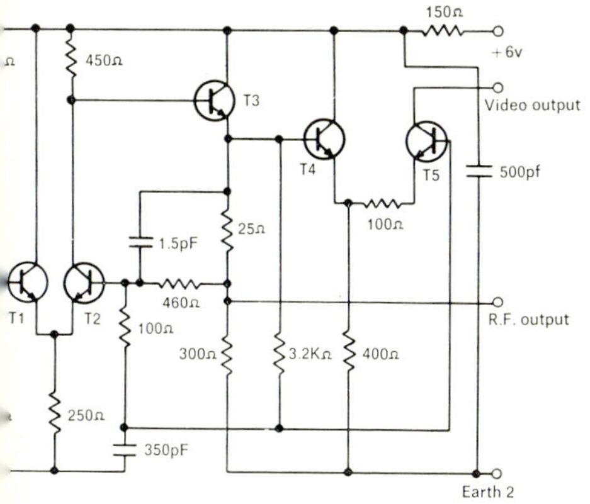

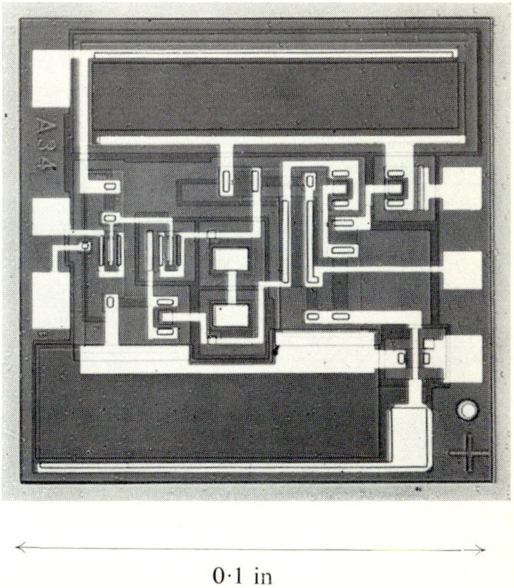

0·1 in

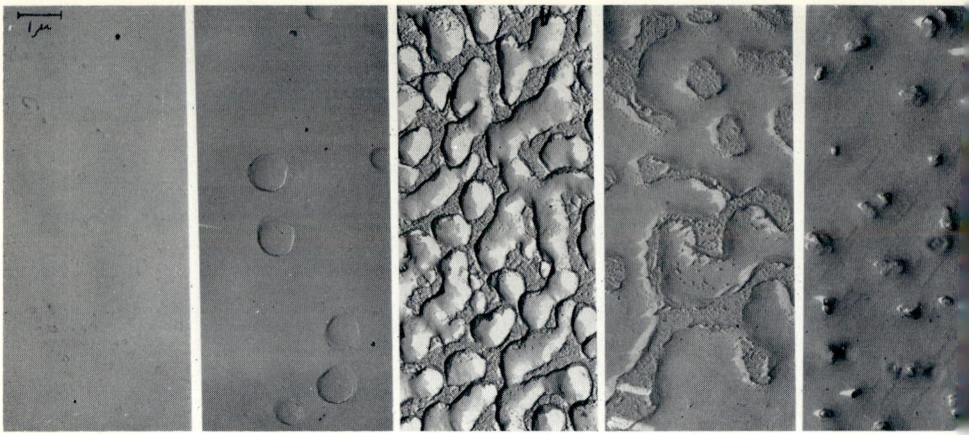

(a) (b) (c) (d) (e)

PLATE 23 Replica micrographs showing growth sequence for silane-grown films ((a) shows alumina surface prior to growth). Minimum continuous film thickness (e) $\simeq$ 2500 Å (*Courtesy Taylor and Francis*)

PLATE 24 Replica micrograph of (11·2) alumina surface after removal of silicon grown from trichlorosilane (*Courtesy Taylor and Francis*)

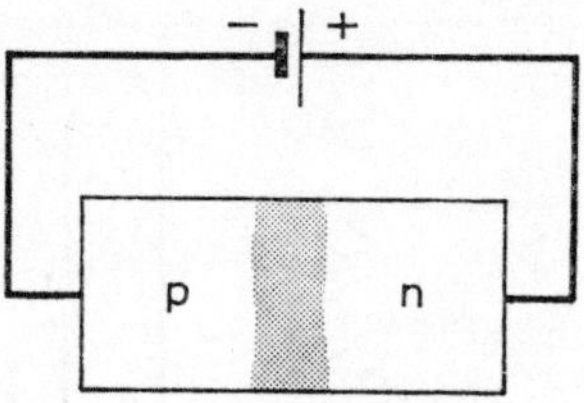

Fig. 45. The p–n junction displaying high resistance

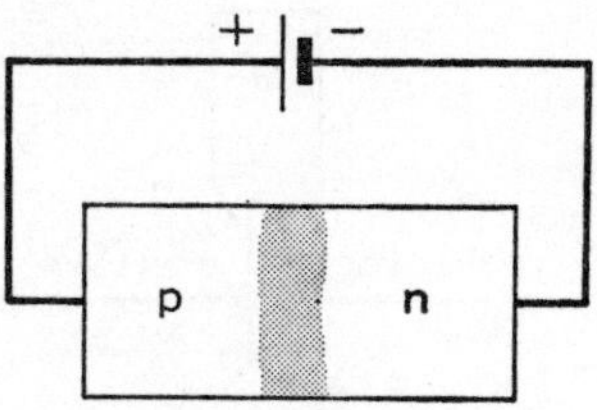

Fig. 46. The p–n junction displaying low resistance

Figures 45 and 46 in fact illustrate a diode rectifier. Such junction diodes are made by diffusing e.g. boron and phosphorus into pure silicon or germanium. Joints therefore do not have to be fabricated between the two regions.

Putting two such devices back to back gives a transistor, which needs no emphasis as to its importance; again, diffusion of impurities eliminates the need for joining (Fig. 47).

Fig. 47. Diagram of a transistor

Bonding however, played some part in the manufacture of the so-called 'Surface barrier' type of transistors where indium or cadmium was plated onto silicon to form simultaneously, terminal connections and a source of regulating electrons (Fig. 48).

Figure 48 shows the device in section and plan, plating being carried out electrolytically by jet with the silicon brightly illuminated to make it conducting.

In this case a satisfactory bond is achieved by electrolytic deposition of metal from ionic solution onto a base which is made temporarily conducting. A reliable bond is dependent on close control of chemical and electrical variables and especially the geometry of the silicon slice.

The most satisfactory components however, have been built up by operating on *one* face of the silicon only (Figs. 49 and 50). Resistors,

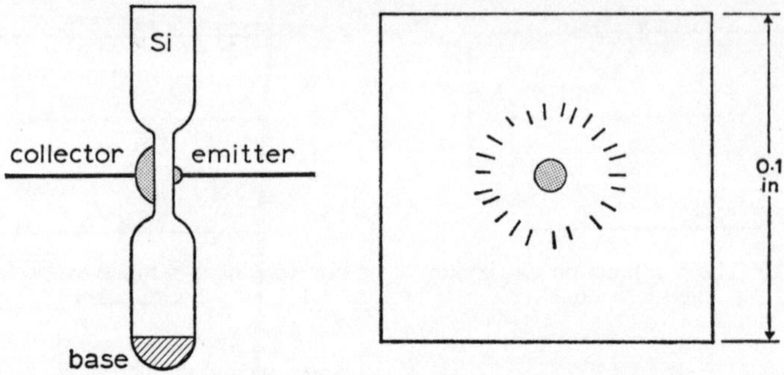

Fig. 48. Diagram of a surface barrier transistor

capacitors and transistors can be made simply by suitable photo-engraving, vapour deposition, diffusion or chemical conversion techniques.

Adhesion between photographic resist, silicon oxide, silica and aluminium or gold conductors is good, providing exceptional care and cleanliness can be maintained.

The complexity of the operations is shown by Plate 22.

Yet with a total area of about 0·1in and correspondingly thin layers of adhering material, dimensional stresses are unlikely to be significant which makes for reliability in construction.

In the manufacturing process silicon is oxidised with steam to SiO_2 and it is probable that compatibility is increased by the occurrence of lower oxides of silica nearer the crystal. Indeed, monoxide is grown on the silicon base to be used as the dielectric in surface capacitors.

It has been mentioned that 'doping' of silicon is carried out by exposing it to gaseous 'p' or 'n' impurities under controlled temperature and pressure (Fig. 51). Now diffusion tends to be rather slow and may be difficult to control in depth as every subsequent heating operation affects previously diffused material.

It has been discovered that single crystal silicon may be used as the basis for a further growth stage, including simultaneously the desirable impurities. This epitaxial crystal development extends the ordered base structure (Fig. 52).

The immaculate character of the original material must be retained for satisfactory electrical performance and whilst about one 'p' or 'n' controlled impurity atom in 10^6 of silicon is required, other foreign

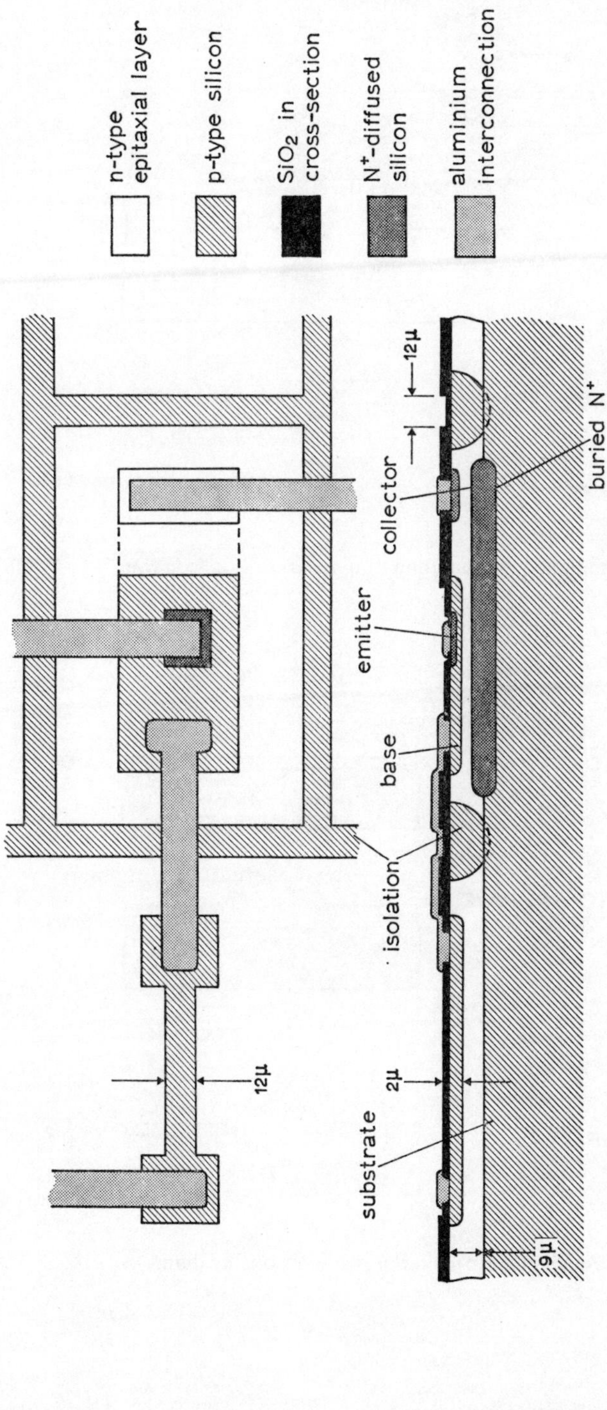

Fig. 49. Diagrammatic representation of a silicon integrated circuit (SIC) resistor and transistor

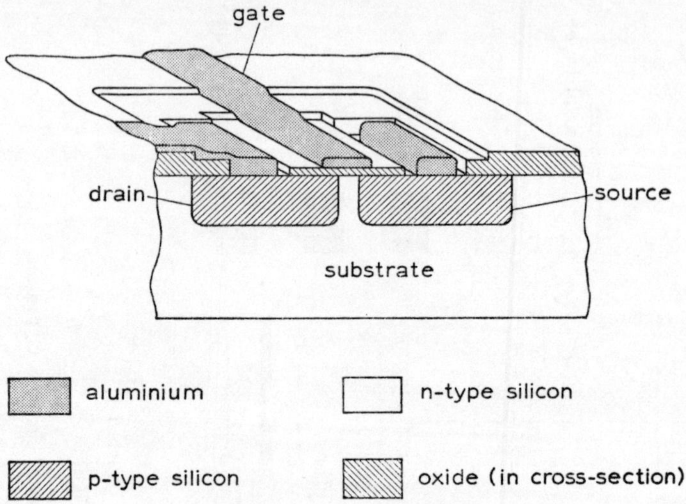

Fig. 50. Diagrammatic representation of a metal oxide silicon transistor (MOST)

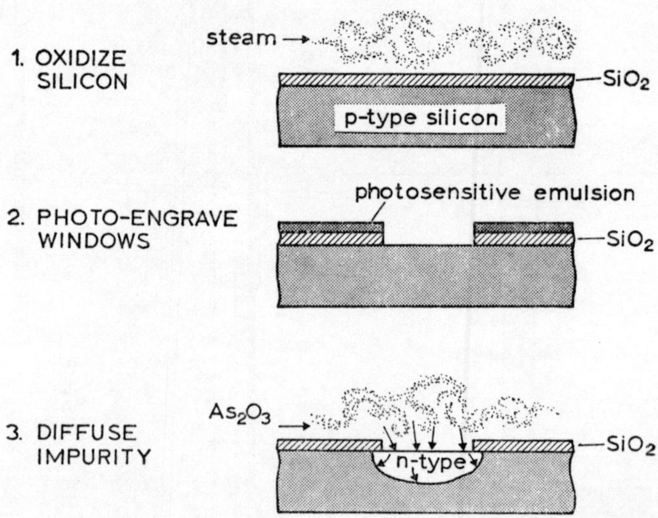

Fig. 51. Construction of p–n junctions by diffusion

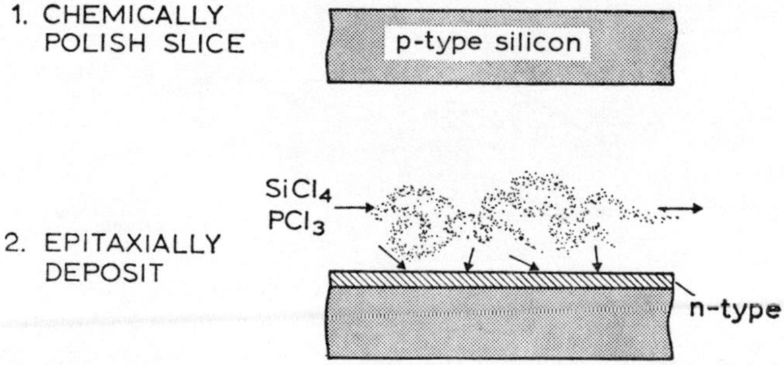

Fig. 52. Construction of p–n junctions by epitaxial deposition

material must be rigidly excluded. Carbon from pump oils will easily react to give silicon carbide and this upsets the crystalline pattern and electrical performance.

Where silicon vapour is deposited on crystalline silicon at 900°C and 10^{-8} to 10^{-9} torr after a pre-treatment including etching with a hydrogen fluoride/nitric/acetic mixture, it is likely that pure covalent bonds are produced. The strength of the surface material is indistinguishable from the original.

Studies have also been carried out by Joyce with silicon deposited on alumina slices cut on the $(1, \bar{1}, 2)$ axis and polished with hot H_3PO_4.[10]

Plate 23 shows stages in the growth and this is compared with Plate 24 where, when the deposited silicon is etched away, reaction is seen to have occurred with the substrate. A schematic reaction might be:

$$Si + Al_2O_3 \rightarrow Si\,O + Al_2O + Al \uparrow$$

and when volatiles can escape the equilibrium is driven to the right.

Though some misfit is inevitable the interface bonding appears to be covalent, probably including highly directional hybridised orbitals and occurs partly because of the growth process which is a heterogeneous reaction involving a chemisorption stage.

With the nearly perfect grade silica surfaces presented by semiconductors, growth centres where attractive forces are greatest are probably located at 'steps' or dislocations in the base material (Fig. 53).

It has been pointed out (Cottrel) by Hellwege[11] that in the structure shown (Fig. 53) attractive forces are greatest at face 1.

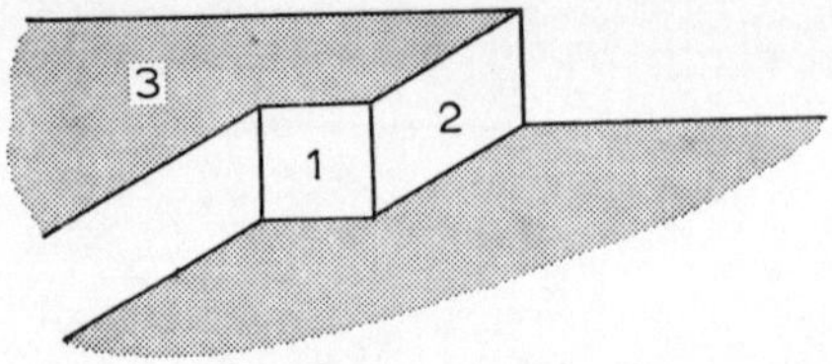

Fig. 53. Diagram of growing crystal

Vol'Kenshtein[12] has discussed fields between semiconductor and adhesion theory. It is worth recalling what may happen to a molecule approaching a solid surface:
(1) It may be affected by van der Waals' forces at 2 to 100Å distance.
(2) It may approach close enough and with sufficient energy for electron exchange to occur, i.e. covalent bonding is possible.
(3) It may bounce off.
(4) If it is adsorbed by covalent bonding it may move over the surface or diffuse.
(5) There may be many centres of adsorption or few.
(6) Adsorbed molecules may interact.
(7) Adsorption is affected by temperature and pressure.

Now $\frac{dN}{dt} \propto N^* s\, \alpha\, KP$

where N is the number of adsorbed molecules of a given kind
N^* is the total number of adsorption centres per unit area
P is gas pressure
s is the effective surface area of the adsorbed molecule.
When κ is constant adsorption is 'inactivated', but when κ varies as $\exp \frac{-E}{RT}$, this indicates 'activated' adsorption,

$$\text{and } \alpha = \frac{1}{\sqrt{2\pi MkT}}$$

where M is the mass of adsorbed molecule
T is absolute temperature
and k is the Boltzmann constant.

It is interesting to note that after the adsorption of the initial molecular cover successive layers on a crystal may tend to counteract the force exerted by the base material. It would seem by inference therefore, that providing contaminants in quantity were absent from a surface, the ability of an adhesive to bond to it would be dictated by its own rather than the characteristics of the adherend.

As we have noted, semiconductor theory predicts diffusion of electrons across barriers between two dissimilar materials such as a metal and a polymer.[13]

Not surprisingly therefore, joints which are broken under controlled circumstances lead to an electrical discharge.[14] An attempt to corrclate adhesive retaining forces with electrical parameters by the simple equation:

$$f = n_o kT$$

where n_0 is a complex involving electronic work functions and charge density in the adhesive at the interface, shows possibilities for further development.

The electronic theory of adhesion has been developed notably by Deryaguin.[15]

Assuming σ is charge/unit area of a metal in contact with a semiconductor, the mutual attraction is given by:

$$F = 2\pi\ \sigma^2$$

This is due to diffusion of charges and the establishment of an opposing electrical double layer at the interface. σ is typically about 10^3 e.s.u. and can account therefore, for at least part of the experimentally measured strength of a joint.

As stresses are applied and strains occur σ decreases due to electron tunnelling up to separation distances of about 15–20Å. Above this the double layer charge density and field intensity in the gap remain constant and tunnelling ceases. (The state of the surfaces may however, cause an opposite effect.)

Beyond 10^{-7} cm where leakage due to tunnelling stops, discharge and secondary emission can occur and voltages of up to 10 kV have been detected at the metal/adhesive interface. Skinner[13] had earlier noted discharges as adhesives were stripped from insulators indicating the presence of such potentials.

One straightforward conclusion (related to the inverse variation of capacity of a condenser with dielectric thickness, mentioned earlier in the paper), is that the thinner the adhesive layer the greater is the electronic contribution to joint strength, a prediction generally realised in practice.

Deryaguin emphasises that the theory obtains only where material diffusion is absent, which indeed rarely occurs. But since E.S. forces are not greatly distance dependent, they are effective at stopping microcracks and especially fatigue. Studies to increase donor/acceptor attraction should lead to more reliable adhesives and composites.

Finally a connection is noted between chemical and electronic theories where, e.g., by chemistry alone it can be imagined that universal adhesives contain different groups capable of associating with partners of opposing polarity on adherends. Using the idea of diffusion of charges between donor and acceptor centres it is possible to arrive at similar conclusions and imagine formation of a double layer and adhesion due to donor/acceptor bonds as the full macroscopic analogue of a heteropolar chemical bond.

Clearly there is a vast and fruitful area existing between adhesion and electronics, studies in which are bound to improve the technologies of both.

References

1. Unreported Plessey work on phenolics.
2. Morris, A. C. and Richardson, A., Environmental resistance of polypropylene, *Brit. Plas.*, p. 85, May, 1968.
3. Benjamin, P. and Weaver C., *Proc. Roy. Soc.*, A254, 163, 1960.
4. Weaver, C., *Adhesion: Fundamentals and Practice*, Ministry of Technology, Maclaren, London, 1969.
5. Hockney, J. A., The Surface Properties of Silica Powders, *Chem. and Ind.*, p. 57., Jan. 9, 1965.
6. Landau, M., Molyneux A. and Zadora-Chrzastowski, W., *Chem. and Ind.*, p. 1949, Nov. 21, 1964.
7. Krotova, N. A. and Morozova, L. P., Application of Infra-red Spectroscopy Methods for the study of Adhesive Underlayer (Polymer-glass) Interaction, *Issledovaniya v Oblasti Poverkhnostnykh Sil, Sbornik Dokladov na Konferentsii Po Poverkhnostnym Silam, Aprel' 1960*, pp. 11–16, 48–54, 83–88, 197–231, Izdatel'stov Akademii Nauk SSSR, Moskva, 1961.
8. Bell Systems Technical Monograph 4552.
9. Whitfield, J. and Cubbin, A. J., *A.T.E. Journal*, **21**, (1), pp. 2–10, 1965.
10. Bicknell, R. W. Joyce, B. A., Neave, J. H. and Smith, G. V., *Phil. Mag.*, **14**, (27), 31, 1966.
11. Hellwege, K. H., *Kunstoffe*, **50**, (1), 5, 1960.
12. Vol'Kenshtein, F. F., *The Electronic Theory of Catalysis on Semiconductors*, Pergamon Press, Oxford.
13. Skinner, S. M., *J. Appl. Physics*, **26**, 498 and 509, 1955.
14. Skinner, S. M. and Gaynor, J., *Plas. Tech.*, p. 626, Nov., 1966.
15. Deryaguin, B. V. and Smilga, V. P., *Adhesion: Fundamentals and Practice*, Ministry of Technology, Maclaren, London, 1969.

Testing High-Performance Adhesives used in Footwear Manufacture

E. F. HALL, F.B.S.I.

THE SHOE AND ALLIED TRADES RESEARCH ASSOCIATION

Introductory

Adhesives are extensively used in shoemaking today. In the United Kingdom approximately 90 per cent of shoes have the soles attached with adhesive. As it has often been stated by adhesive manufacturers serving the shoe and other industries that the high performance requirements of a sole attaching adhesive are sometimes difficult to achieve this talk will be confined to this aspect of shoe assembly. The problem of achieving a balance between the optimum production and service requirements is complicated by a demand for an ever-increasing production rate of shoes and the variety of materials (which are to some extent subject to change, as fashion dictates) which the adhesive must satisfy. Let us first look at the materials that are used.

TABLE 15

UPPER MATERIALS

Leather	(full chrome) (semi chrome) (vegetable tanned)
Poromeric *	e.g. Corfam
Special finishes	e.g. coated leather or fabric (1) Nitrocellulose (2) PVC (3) Polyurethane
Plain Fabrics	Natural and synthetic fibres
Mesh Fabrics	Natural and synthetic fibres
Pile Fabrics	e.g. velvets, corduroys, suedenes

*Often referred to as coriaceous (leather-like) materials. There are however, a number of excellent leather-like upper materials now available which are not permeable to water vapour. Poromeric are permeable.

TABLE 15 (continued)

SOLING MATERIALS

(1) Leather
(2) Composition rubber
(3) Resin rubber
(4) Microcellular rubber
(5) Translucent rubber
(6) Plastic PVC†, Nylon‡
(7) Unvulcanised rubber (crepe)

†Use is increasing rapidly ‡Only on very limited scale

TABLE 16

ADHESIVE

Type	*Comments*
Cellulose nitrate (pryoxylin)	Was extensively used on leather but is not suitable for streamline shoe production
Natural rubber (non vulcanising) latex and solvent solutions	Confined to crepe rubber but providing primer and high pressure is used neoprene adhesive is more satisfactory
Heat curing natural and synthetic rubber solution	Used for the rubber moulded-on process
Nitrile (hycar)	Hardly used today
Neoprene	Extensively used single and two component. PVC surface must be primed
Polyurethane	Mainly for PVC substrates although now cheaper and superseding neoprene
Hot melt (polyester or ethylene copolymer bases)	One-way system, adhesive applied to sole only

Note: Two-part adhesives are used on a limited scale but are not normally liked because of the difficulty of machine application and the uncertainty of hand mixing. (Urethanes are the exception, as these, even when well mixed, can have a very useful pot-life.)

A few comments on these materials in respect to their bonding properties may be helpful.

Leather is still the more predominant and to a large extent the more pre-eminent upper material. It has, and still does give some of the more baffling problems of adhesion but more recently these have been confined to trouble associated with the processing greases and not to the structure. For example, the curing properties of natural rubber adhesive can be retarded or inhibited by the acidity of a leather, or the fatty acid content of the grease can be troublesome with neoprene adhesives. However, the layerwise physical structure of the leather is very important as we shall see when considering methods of preparation. Of the three leathers listed in Table 15 the full-chrome usually gives the strongest bonds and semi-chrome upper leather can be the most troublesome.

With the type of adhesive used on certain synthetic upper materials e.g. nylon, polypropylene and polythene, mechanical keying is essential to obtain a bond adequate to hold even the lightest of soles to evening shoes. Just recently we have been approached by a Dutch shoemaking member to solve the problem of sticking a sole to polypropylene straw fabric (large orders of this shoe had been taken and enough material to make them had been acquired before it was found that the sole could not be stuck to the upper). Fortunately in this case, the straw fabric was laced with strips of leather which when roughed at the same time as the polypropylene surface, was sufficient to provide enough anchorage for the adhesive. Fabric meshes, the type of upper material often seen on ladies' evening shoes, is another example when mechanical keying into the interstices plays an essential role but even these delicate materials are usually lightly roughened to teasel up the surface.

It can, therefore, be seen that surface preparation is most important and the principal methods used will be featured throughout this talk.

Production technique

Before outlining the methods used to evaluate a shoe adhesive a brief description of the basic construction and production technique used for sole attachment will clarify the meaning of some of the test methods used.

Figure 54 shows the most common construction where the sole is bonded to the upper material which is lasted in over an insole. The sole can be cut from leather, plastic or rubber sheet or pre-moulded from PVC or rubber. This is known as the stuck-on or cemented method. It can also be moulded to shape in situ by compression moulding rubber or injection moulding plasticised PVC (or in a few cases rubber).

Surface Preparation

As with most adhesive bonding the art begins with surface preparation. Mechanical methods are almost universal but solvent

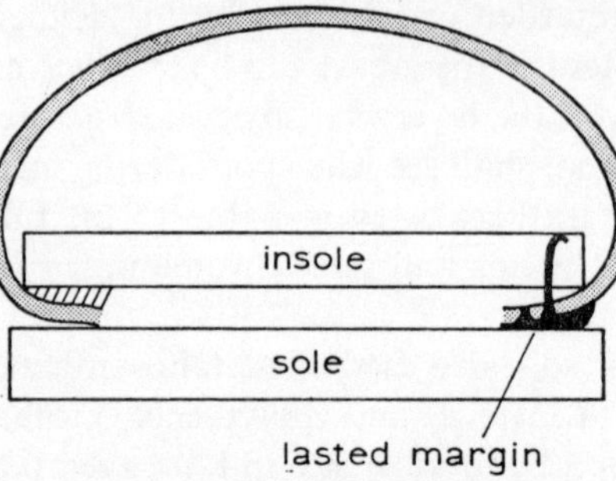

Fig. 54. Diagram of a cross-section of a 'stuck-on' or 'moulded-on' shoe where the upper is lasted and attached to an insole. The lasting operation involves pulling the upper material over a wooden last and tacking or cementing the margin to the underside of an insole. The outsole is then attached to the turned in margin by either one of the following processes. (1) Sole prefabricated from sheet or pre-moulded to shape and attached with adhesive. (2) Rubber sole compression-moulded in situ pre-applying adhesive to the margin of the upper. (3) PVC or rubber injection moulded in situ, pre-applying adhesive to the upper.

wiping or priming is being more extensively used on plastic and rubber materials. Although roughing will increase the surface area, breaking up the surface of some materials, a low grade resin-rubber for example, can reduce the bond strength. Upper materials, such as cellular PVC-coated cloth, should be solvent wiped, as damage to the surface skin can cause a weak link as it will initiate a tear more easily into the weak sponge layer, see Fig. 55(a).

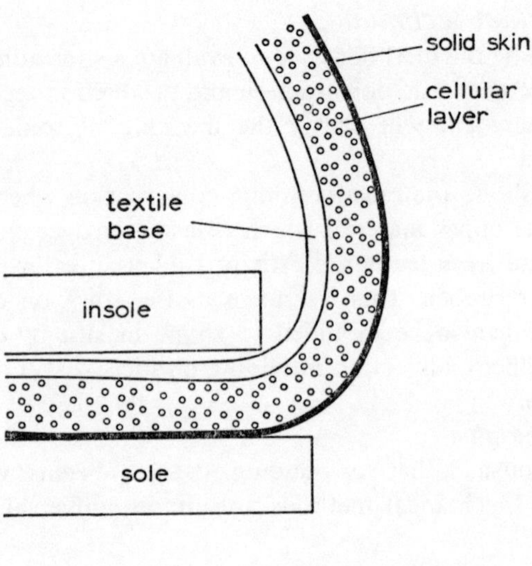

Fig. 55 (a)

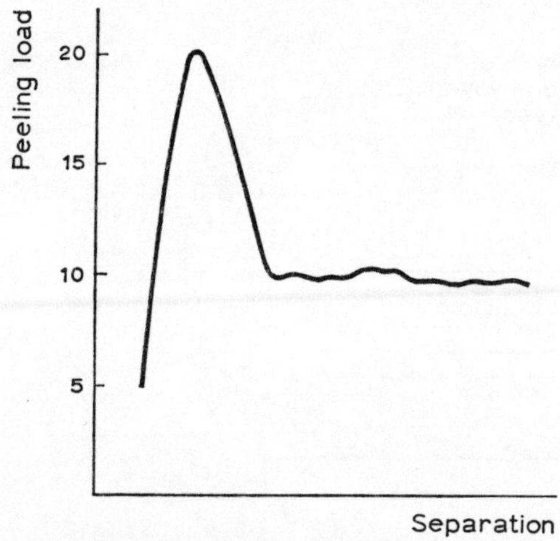

Fig. 55 (b). Cellular PVC coatings, often on a knitted base, have a solid surface skin which must be strong enough to support a stuck-on sole. Mechanical roughing can weaken the skin. The peeling curve shown is typical. The drop from 20 lb/in^2 width to half this value or even less is caused by the solid skin breaking through and separation progressing in the cellular layer. If the surface is roughed the initial load, which determines the strength of sole attachment is reduced.

Poromeric (e.g. Corfam) upper materials (Fig. 56(a)) should be only lightly roughed as the inter and base materials can give a weaker foundation than the urethane surface layer.

Obviously mechanical roughing is undesirable on any fibrous material if done excessively, lighter upper materials can be seriously weakened and will tear just above the sole junction. On the other hand if solvent wiping is used contamination of the finish on certain upper materials can be most troublesome.

Any chemical treatment of the lasted margin, e.g. oxidising, is not suitable because of the danger of damaging or marking the upper. Also the time required for doing this on the making track could not be allowed.

Adequate adhesion can be achieved without preparation on certain PVC and poromeric upper material, and this is the characteristic that has been sought after for many years.

Unfortunately leather, especially finished grain leather, must normally be well wire brush roughed to remove the finish and usually the weak grain layer (Fig. 56(b)). This outer layer can also contain a high concentration of grease which can, if the fatty acid content is

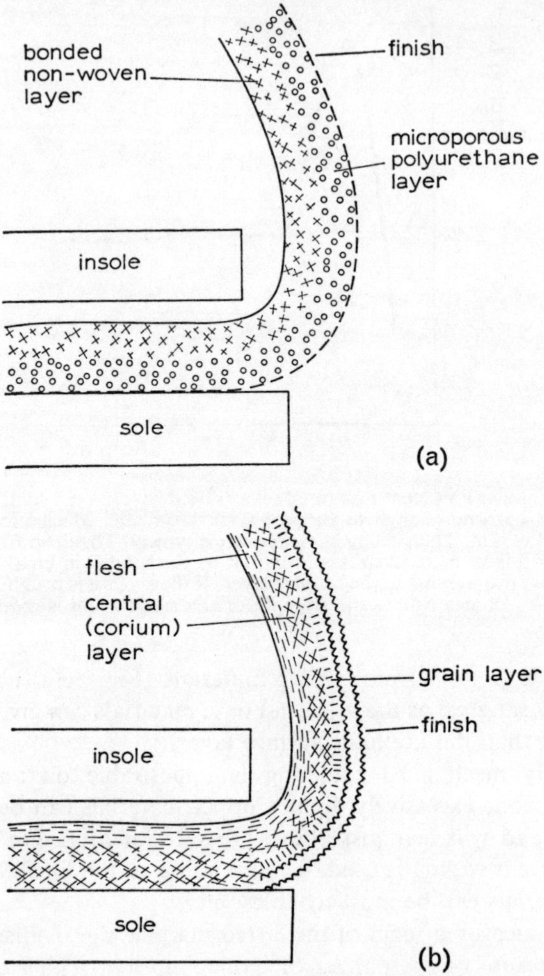

Fig. 56. (a) Attaching a sole to most poromeric man-made upper materials requires the surface to be only lightly roughed. (b) It is essential to remove the finish and normally the grain layer on upper leathers.

high, migrate into the neoprene and react with the stabilisers to form insoluble salts (metal soaps) which can impair coalescence of two neoprene films. Grease can also prevent complete coalescence of hot PVC to certain urethane adhesives in the injection moulded-on process. In this case the grease reacts with the isocyanate cross-linking agent if present.

TABLE 17

Normal methods of preparing soling materials

Soling	*Preparation*	*Comments*
Leather	Rough	Weak flesh fibres must be removed
Resin-rubber (high styrene butadiene-styrene co-polymer)	Split by band knife or scour	Roughing can weaken the surface
Translucent rubber (silica filler)	Rough and Lacsol * wipe	Very tough rubber difficult to remove and obtain entire virgin surface
PVC	Scoured or solvent wipe	Scouring is preferred as solvent wipe cannot be seen when dry
Nylon	Prime with solution of nylon in formic acid	Special care must be taken in use

*Solvent based acid treatment which combats zinc stearate bloom left on the surface, by converting into stearic acid.

Application of adhesive

One coat of adhesive is invariably machine applied to each adherend. One-way adhesives, mainly of the hot melt type applied to the sole only, are being used on a limited scale.

Method of assembly

Although likely to change over the next few years the production methods used require the treated soles to be stored several days or even weeks before joint assembly. The applied adhesive must be reactivatable to revive sufficient surface tack immediately before being assembled to the shoe upper. Solvent activation was used with cellulose nitrate adhesive but demands excessive press times. Heat activation is now almost universal but softening of the synthetic rubber based adhesives, and more recently the hot melts, under intense heaters for extremely short exposure times has provoked some interesting problems in respect to the uniformity of heating and temperature measurement.

The 'dry tack' produced by heat softening must be sufficient to hold the sole to the shoe upper, after first spotting in position by hand, and during transfer of the shoe to the press. An almost instant bond is required after consolidation in the press to a standard that will resist the separating forces produced by the curved sole trying to revert back to a flat shape. (If the curvature of the shoe bottom is excessive such as on a women's high heel shoe the sole is sometimes pre-conformed to shape). The time in the press on modern production lines, producing stuck-on shoes, can be no more than ten seconds.

With the moulding-on methods the time in the moulding machine is the 'closed assembly' time. This can be five minutes or more for vulcanised rubber and less than a minute with PVC injection.

Testing

Type and preparation of test specimen

With most of our adhesive work there is a certain lack of reproducibility due to the use of a biological material, e.g. leather, in several of its different forms. If, on the other hand, testing was confined to more uniform materials e.g. textiles, considerable information on the effects of the adherend on the bond strength would be lost. (Most of the recent troubles in practice have been due to substances migrating from leather and rubber adherends.) Standards of most materials given in Table 15 are held in the laboratory for test work.

Preparation of the surface

These are the same as previously described under factory procedure. The type of roughing or scouring tool, the speed at which it rotates and the frequency of changing or sharpening are factors that must be well controlled. The procedure of cleaning surfaces e.g. PVC are well established.

Application of adhesives

The application of a uniform coat of known amount is most difficult by hand methods as was found in the early stages of our work when much attention was given to methods of applying known weights or volumes. All methods presented difficulties and, although the assembly of larger test blocks overcame some of the problems, the gain achieved in improved reproducibility did not appear to justify the increased use of expensive substrates. Careful brush application, applying one or two coats, is normally carried out.

Room conditions

The effect of extreme room condition when preparing and testing joints has not been as significant as was expected.

High humidity when preparing the specimen can influence the results and conditioning before and after assembly is to be recommended. However, the unpleasant working conditions imposed when applying the adhesive in a controlled atmosphere, unless very expensive conditioning plant and pleasant artificial light is utilised, do not appear to be justified.

Evaluating certain working properties

Tack and green strength

Of the tests performed those determining tack and green strength deserve special mention.

As stated before, the adhesive must hold the sole to the shoe immediately after being positioned under light finger pressure. The property required is usually referred to as spotting tack and has been roughly assessed by assembling the joint under light loads and carefully separating it by hand. On the basis of this work a tack tester is now being constructed. Compared with neoprene adhesives the more recently introduced urethane solutions often have poor tack under similar conditions of open time and heat reactivation.

The nature of the green strength test is probably a little more unusual. This, as the name implies, determines the initial bond, in fact the strength as soon as is practicably possible after the joint is released from the press.

The test is extended to also determine the rate of set-up, as the speed at which the bond strength develops after assembly governs the time interval between this and the next operation (normally today only a cleaning job, which on some track conveyors is required in a minute, or so, after sole assembly).

Should the sole bond not remain tight to the upper because of slight creep or 'springing' early failure in wear can be expected due to cohesive failure of the adhesive. The ability to achieve a tight bond quickly is, therefore, of paramount importance.

The residual stresses in the freshly made joint will depend on the physical characteristics of the soling material (modulus and thickness) the curvature and probably above all on the amount of stretch applied to the sole during attachment. An extreme example of the latter would be longitudinal stretching that would be required to fit a sole that is half a size shorter than the upper (see Fig. 57).

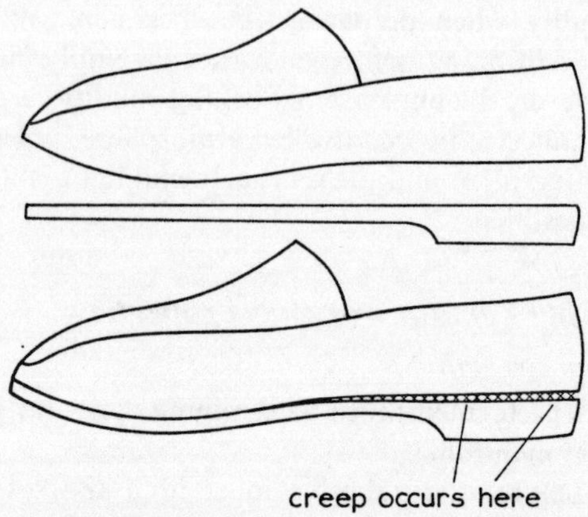

Fig. 57. When a sole is fitted to a shoe upper, residual stress in the freshly made joint can be set up, especially in the waist and heel regions. It is common practice to stretch the sole slightly when fitting it to the upper. The ability of the adhesive to resist these residual forces is very important and the property is defined as 'green strength'

Solvent retention in the applied adhesive film (Fig. 58) or excessive heating of a thermoplastic adhesive, especially when applied to a stiff leather sole, can also cause this trouble. It is, however, one of the few faults that can, providing the shoe is carefully examined, be seen before the shoe leaves the factory.

A slow speed peeling test (jaw separation 1/20 in per minute) is carried out to determine the green strength and the rate of build up of strength in a machine that has the essential feature of recording a load as quickly as possible after assembly.

A small pneumatic press (Plate 25) that can apply 80 lb/in^2 pressure forms part of one of Satra's peeling tensiometers. This allows the joint to be assembled whilst in the loading position on the machine (the ends of the components forming the test pieces are clamped in the jaws of the machine before assembly and before heat activation if this is done). During the 10 to 15 sec dwell in the press the test piece is clamped to the winch and recording mechanism. The winch can be manually turned by a handle to load the specimen very quickly immediately the joint is released from the press. A load in excess of 15 lb is applied in doing this and then the handle is released so that

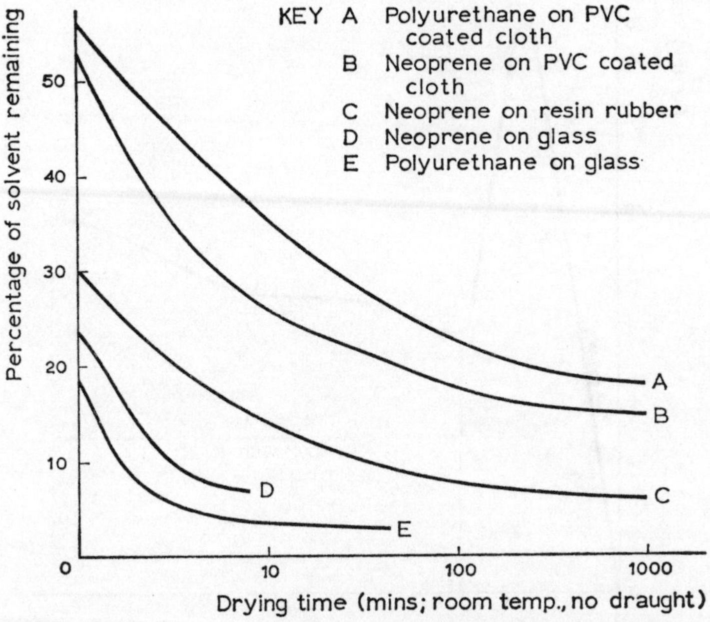

Fig. 58. Solvent retention in the applied adhesive can be most troublesome as most shoe factories rely on rapid force-drying. Apart from temperature and air speed the drying rate can be affected by the type of substrate and amount of adhesive applied

separation will continue at the much slower rate of 1/20 in per minute. By selecting an initial loading of 15 lb a drop in load occurs at the much slower speed of separation and the lowest point on the graph is recorded as the 'immediate strength'.

The time taken to build up to 10 lb per inch width from this point with the joint separating at 1/20 in per minute is further recorded (see Fig. 59).

Stresses involved in wear

The stresses that can occur in a sole to upper bond are complex and the mechanics involved are not fully understood. However, the principle forces (see Fig. 60) that can cause breakdown are:

(1) Impact, e.g. stubbing the toe of the sole against the edge of the pavement.

(2) Peel, resulting, for instance, from the upward pull of the foot on the upper.

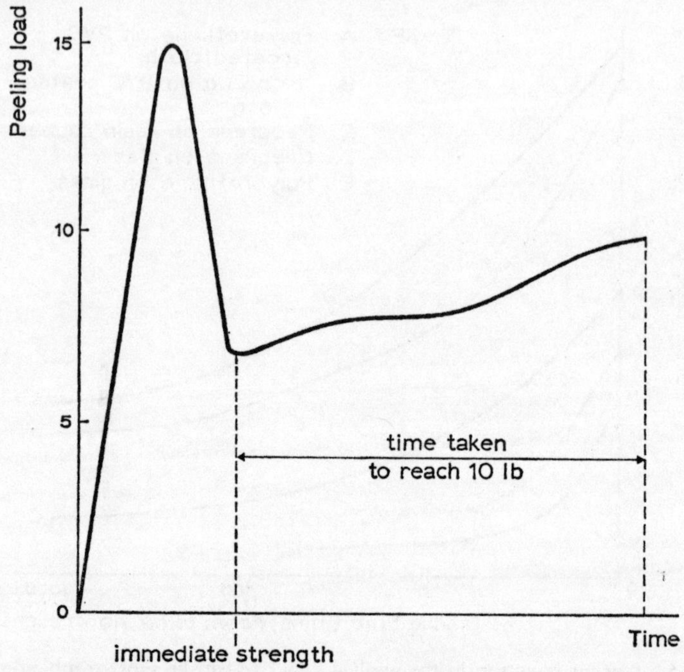

Fig. 59. By quickly turning the handle of the testing machine the 1 in test specimen is quickly loaded to 15 lb. The manual operation is then discontinued and the motor drive at a very slow speed separates the joint at 1/20 in per minute. The load normally drops when changing from manual to motor drive and the lowest point on the graph is recorded as immediate strength. The time taken to build up to 10 lb is then recorded

(3) Shear, resulting from the tendency for the insole and upper to resist shortening and the sole to resist lengthening, when the shoe is bent in walking.

Impact at the toe can cause high peeling forces if the upper and insole forms a rigid structure and the sole is very flexible. Breakdown can occur by one large force or repetition of a smaller force. As a rule the upper is much more flexible than the sole and if an upward force is applied to the sides of the upper it will tend to peel away from the sole. The insole which is normally much more rigid will have a marked washer effect in spreading the load.

(It is for this reason that a mocassin shoe, which has no insole, will often fail at the tread position if the sole is rather rigid, e.g. leather). The forces tending to break the adhesive bond in the tread

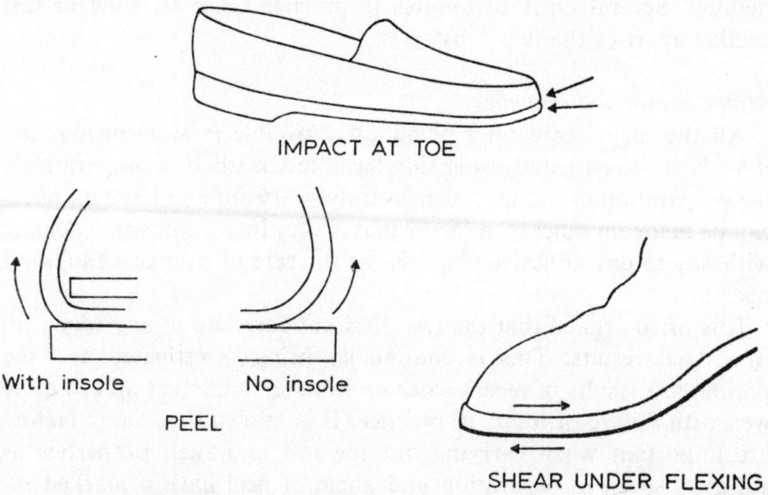

Fig. 60. Peel and shear stresses can occur in a sole joint. High rates of peel can occur under impact at the toe and a peeling action can occur from the upward pull of the foot on the upper. Cyclic loading obviously raises the need for a dynamic test but the evolution of the most suitable form of test has taken many years to complete

position of the shoe, when it is flexed, are largely shear stresses and are highest when the sole is stiffest and thickest and when the insole/upper structure is also stiff.

Performance tests

Impact strength

Several attempts have been made at establishing a suitable impact test. In the early stages of our work we tried tearing apart an overlap joint with the B.C.I.R.A. Ballistic Tester, used for determining the energy absorbed in tearing textile specimens.

The results achieved were very much influenced by the physical properties of the substrate and the striking position of the pendulum hammer on the joint.

Peel test

Although this test has been criticised on the meaning of the measurement, e.g. it is a long way from measuring true interfacial adhesion and it is affected by so many variables, it gives results that can be readily understood in the practical use of adhesives in shoemaking. Several types of failures in practice occur by slow or fast peeling apart of the two substrates.

Major causes and remedies

All the major epidemics of adhesive trouble in shoe production have been investigated using this basic test method. Comparatively the preparation of the specimen is straightforward and large numbers can be made up quickly, a factor that is very important when dealing with day to day adhesive enquiries at the rate of over one thousand per year.

It is often argued that the test does not correlate in any way with wear trial results. This is continually being investigated and the peeling test results of recent work on hot melt adhesives agreed quite well with failure of joints in practice. It is known that many factors are important when carrying out the test and such properties as stiffness, speed of separation and angle of peel have a marked influence on the results obtained.

Without going deeply into detail of the peel test method used it can be stated that the attainment of consistent and uniform heat activation during the assembly of the test piece has involved much study, and evaluating methods of temperature measurement has been fundamental to it. It has been found that the crystals of pure organic chemicals are more effective than heat sensitive paints and a very fine thermocouple connected to a quick response meter is the best laboratory method.

Pneumatic presses are used to consolidate the bond (hand rollers, etc., have not been reliable enough) but the bond strength does not appear to significantly improve above the pressure that is required to bring the surfaces closely together. Pressures within the range of 40 to 80 lb/in^2 are typical in the laboratory and factory. High pressures can excessively compress materials such as microcellular rubber and cause higher than normal residual stress in the joint.

A small daylight press and a suitable flat mould is used for bonding vulcanizable rubber to upper materials. Likewise, by using a small injection moulding machine plasticized PVC is moulded to adhesive-treated materials.

Where possible the average load is estimated from the autographic recordings but often the lower loads and the frequency they occur are assessed on the grounds that failure will occur in practice where the joint is weakest. Analysis of the type of failure in good and bad joints is essential in all our work and our belief in the importance of this in the factory and laboratory has strengthened over the years. Exercises for identifying faults and analysing the type of failure have formed a prominent feature in some of our Adhesive Seminars for managers and foremen, as a knowledge of the cause and remedy of faults can form an invaluable aid to trouble-shooting in any factory using adhesives intended for permanent assembly.

Reports and manuals containing fault-finding charts have been issued over the years and a specimen page from a recent report is shown below.

NON-COALESCENCE OF TWO ADHESIVE FILMS

A common weakness and bad fault and the cause could be:

(1) Time interval between cementing and assembly too long
(2) Insufficient heat reactivation
(3) Insufficient pressure
(4) Grease in upper leather
(5) Excessive pleats in lasted margin (part adhesion)
(6) Contamination of either adhesive coating
(7) Bottom filler overlapping lasting margin
(8) Upper too large giving poor fit of mock welted unit
(9) Excessive penetration of cement

Of special note (on above)

(1) Still common practice to leave treated soles for several days or even weeks before assembly, then only heat reactivate shoe uppers
(2) Temperature, exposure time and voltage should be periodically checked
(3) Often caused by press pad being incorrect shape
(4) High grease or fatty acid content of upper leather can drastically reduce tack
(5) Usually occurs with tack-lasted work
(6) Contamination from hands etc. (not likely).
(7)
(8) } The treated sole is prevented from contacting upper
(9) Can occur when through cementing leather soles

Specimen page from a report that is intended to be useful as a factory manual. Intelligent analysis of the type of failure that can occur is an excellent aid to trouble-shooting in the factory.

Resistance to environmental conditions

The peeling test piece is exposed to heat, water and oil (usually when ascertaining whether industrial boots with a moulded-on sole will withstand repeated soaking in certain oils) and then tested in the normal way. The test joint is also tested at elevated temperature (60°C) but the sustained load heat resistance test is preferred.

In this test 30 one-inch peeling test pieces are prepared, half are tested after 48 hr and the other half 4 weeks after assembly. The two series of tests determines whether the joint strength has matured (e.g. by curing of the adhesive) or weakened, e.g. due to plasticiser migration, during the time interval. A set of five joints at a time are suspended in an accurately controlled heat cabinet with good positive air circulation. After 30 min conditioning at the temperature required (normally 40 or 60°C) the amount of separation that occurs in 10 min under sustained load of 1 to 5 lb is recorded.

As would be expected, the test is most sensitive to a number of variables and the conclusion from a recent investigation using the test on urethane adhesives showed:

(1) Type and brand of cement had little or no significant effect.*
(2) Open times of cement films can affect the result if too short.
(3) Ageing of joints before test is extremely important.

Care must be exercised when interpreting the results from heat ageing treatments. For example, in recent work on plasticiser PVC/resin rubber joints it was found that neoprene adhesives gained 10 per cent in weight by plasticiser absorption after one year at room temperature and 20 per cent after six days at 50°C (in some cases the figure tended to fall again at 14 days at 50°C). The ageing test at 50°C although probably realistic for shoes being displayed in shop windows or even when being transported to tropical climates could be unfair to neoprene adhesive as an accelerated room temperature test.

Direct tension test

Attempts to overcome the limitations of the peeling test in giving a true measure of adhesion by using a force on the test piece applied at right angles to the plane of the load have not been fully successful in our work. Figure 61 shows the first attempt where circular specimens

*Previous work with neoprenes has shown that leather/leather and leather/resin-rubber do not always give the same answer.

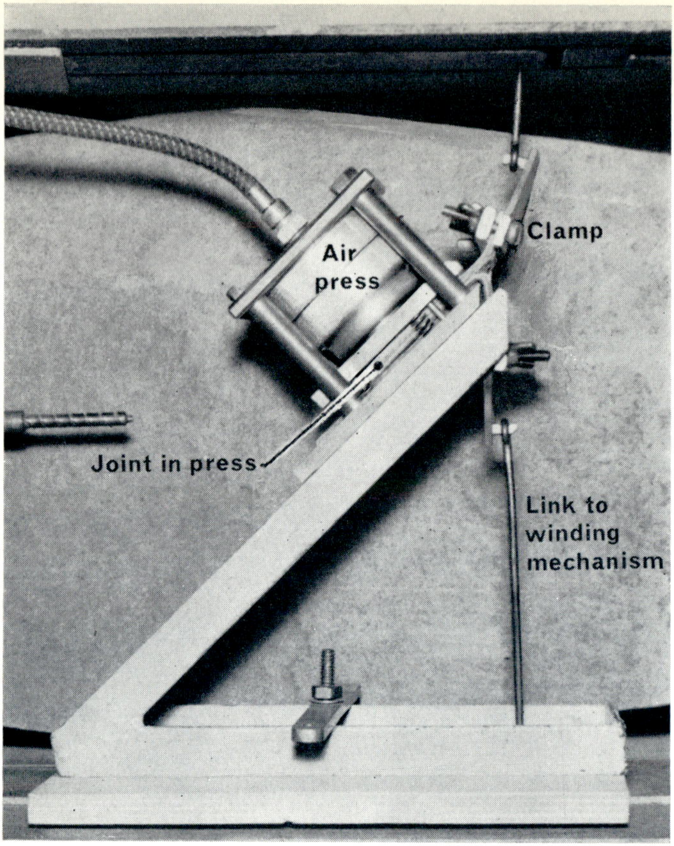

PLATE 25 Apparatus for green-strength measurement. This illustration shows the device which is mounted in the ordinary Satra peeling tensiometer in order to measure the green-strength. The test joint is made in the apparatus itself, being pressed for 15 sec by the small air press shown. Since the ends of the joint have been previously mounted in the jaws of the tensiometer, peeling can begin immediately after the joint has been made

PLATE 26 In this test a lap joint is subjected to repeated stresses as in the lifting weight machine. The weights are replaced by springs

PLATE 27 A joint of flexible upper material bonded to a more rigid soling material is attached by stitching (the stitches passing through the extended ends of the flexible material) to a textile belt running over two pulleys. All dynamic tests are carried out in a thermostatically controlled cabinet

of leather were held in screw clamps which, when tightly closed, raised the surface sufficiently for it to be treated with adhesive and bonded to a similar clamped treated specimen. The stretch of the substrate under strain introduced considerable peeling action which was unwanted behaviour in this case. The well known method developed by Wake for rubber/textile structures has been more useful but the accuracy required for aligning the specimens, the difficulty of bonding certain materials, e.g. PVC, to the metal, the extension that can occur of elastomeric substrates unless very thin and the unwanted failure of the bond attaching the specimens to the rod that sometimes occurs, often prevent ideal measurements. Nevertheless, the test has been found useful in research work.

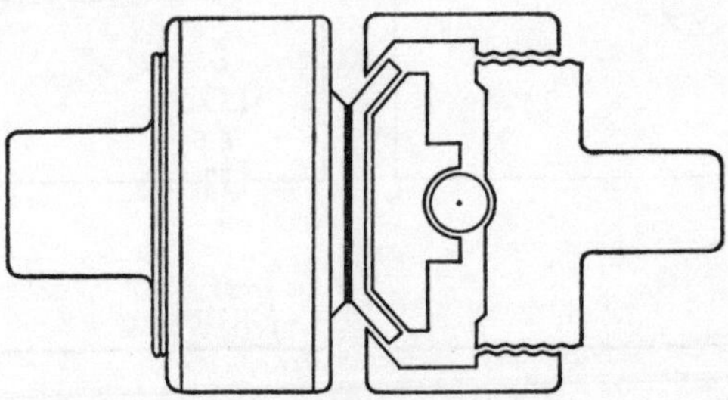

Fig. 61. Tension test clamps that were tried in the early days of evolving Satra's adhesion test methods. It was found that separation of the two clamped materials occurred mainly by peeling action due to the bulging of each flexible substrate under tension. It was also difficult to ensure axial and symmetrical alignment of the two materials when brought together

Fatigue testing

The need for evolving a suitable test for the sole/upper bond under cyclic loading was apparent in the early days of our work. At that time natural rubber adhesives were being used on a limited scale and although the peeling loads obtained were considered to be of the required standard, failure of the bond in wear was most common. Much of the trouble was due to contamination (minute traces of copper in certain forms will have a marked effect) but it was found that comparatively simple fatigue tests could detect a dangerous weakness in joints assembled with non-vulcanizing or partly cured natural rubber solutions. The Hall Flexing Machine (used for evaluating the flex resistance of rubber solings) was the first machine considered. In this test the bonded specimen was bent to form a loop and 'rolled' between the two horizontal plates. The heat 'build-up' in the specimen and a tendency for the specimen to kink in one place were troublesome factors. Another more simple machine for repeated flexing uses a continuous running belt enclosed in a constant temperature cabinet. The test specimens are held on the belt by stitches (Plate 27). More elaborate machines have evolved from the simple principle of repeatedly lifting a weight from the floor (Fig. 62).

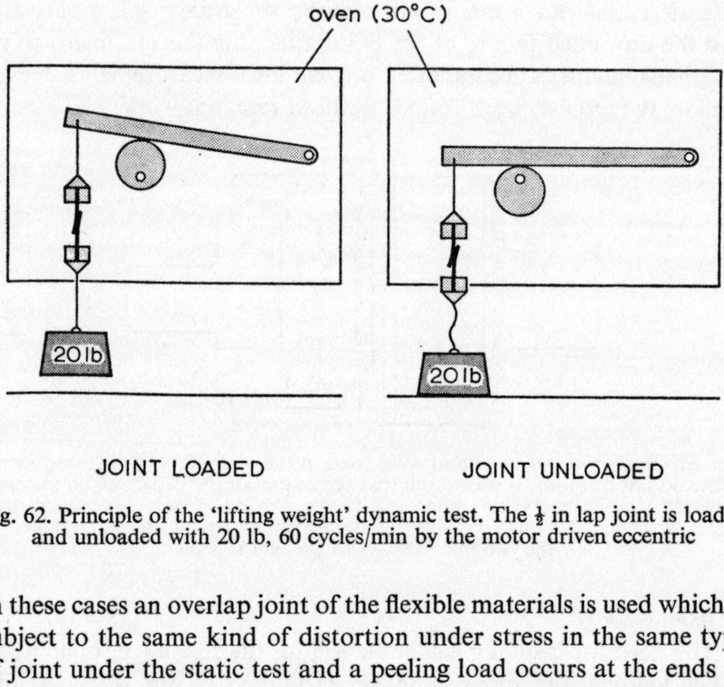

Fig. 62. Principle of the 'lifting weight' dynamic test. The ½ in lap joint is loaded and unloaded with 20 lb, 60 cycles/min by the motor driven eccentric

In these cases an overlap joint of the flexible materials is used which is subject to the same kind of distortion under stress in the same type of joint under the static test and a peeling load occurs at the ends of the bonded area. The test under the dead weight method suffers from appreciable shock stresses and as the specimen extends and separates the time under load progressively decreases. Replacement of the weights with springs (Plate 26) reduces these drawbacks but the use of a load applied by direct pneumatic or hydraulic force is probably the best method that can be achieved when using this type of test piece.

The Effect of Moisture on Adhesive Bonds

N. C. MACDONALD, B.Sc., F.R.I.C.*
A.W.R.E., ALDERMASTON

Introductory

At the present time there are three principal theories of adhesion, the adsorption theory, the electrostatic theory, and the diffusion theory. Each has its own ardent devotees and the literature in recent years has contained several powerful arguments for and against each of these theories.[1, 2, 3] Balanced opinion seems to indicate that there is no overall theory, rather a synthesis of all three applies, the relative importance of each individual theory depending on the chemical natures of particular adhesive/adherend combinations.

This paper is concerned with the behaviour of a particular class of adhesive joints in humid environments. The joints considered are those involving metal adherends bonded with adhesives of the so-called structural variety. Although it cannot be claimed that the survey covers many of the large number of combinations of such polymer and substrate types available, the treatment of the effect and the theoretical principles involved should apply generally.

The whole question is viewed in the context of the adsorption theory, since this appears to be the most important for the case of adhesive joints of the above type. Kiselev's treatment of specific and non-specific interactions of molecules of different electronic structures with solid surfaces[4] lends support for such an approach.

Studies of the effect of moisture on adhesive joints at Aldermaston followed investigations into the behaviour of surface coatings films in wet environments. Several workers had reported earlier that in water-soak conditions paint films could lose an appreciable proportion of their adhesion to substrate: in particular Dunkley and Earp[5] reported that alkyd and oleoresinous paints were affected and James[6] reported similar behaviour with certain epoxide resin based materials. Walker has commented on the paucity of quantitative

*Now at Evode Limited, Stafford

data and in his Roon Award paper to the Federation of Societies for Paint Technology in 1965 he has surveyed the whole problem, including his own work.[7] Consideration of the effect of moisture on structural adhesive joints and the proposed mechanism are a direct extension of Walker's detailed observations with paint films which are summarised in the following section.

The effect of water on the adhesion of surface coatings[7]

A wide, and representative, range of surface coatings materials was examined both in water-soak and saturated water vapour environments, and adhesion to substrate determined by the direct pull-off technique. For the purposes of this paper only the high humidity effects will be dealt with since these are directly comparable with the conditions used in the structural adhesive work.

All of the systems examined showed loss of adhesion in 100 per cent relative humidity (RH) conditions at room temperature. At lower RH values the loss of adhesion effect was less marked, or even absent. Thus an alkyd paint film was largely unaffected until the RH level reached 80–90 per cent; an epoxide ester system showed similar behaviour. Of the two solvent-drying systems investigated, the chlorinated rubber based material was virtually unaffected until the RH rose above 90 per cent, when rapid loss of adhesion occurred. On the other hand a styrene-butadiene copolymer based system began to deteriorate in performance when the RH was 80 per cent. The two chemically cured systems examined, a polyester/isocyanate and a polyamide/epoxide, showed greater sensitivity to humid conditions, the onset of adhesion loss occurring at RH levels in excess of 50–60 per cent. The best performance was achieved with a thermosetting acrylic paint which was stoved at 180°C; in this case loss of adhesion was apparent only in 100 per cent RH conditions, and even so the loss at equilibrium amounted to less than half the original value. Figures 63–66 show some of the decay curves obtained by Walker.

The foregoing results were obtained with stainless steel substrates. The importance of standardisation of surface finish and preparation is emphasised but Walker does point out that a similar pattern of behaviour was observed when aluminium was used as substrate. Not unexpectedly complication of the picture occurred when brass substrates were used, due to side effects such as oxidation of the polymer or corrosion of the substrate.

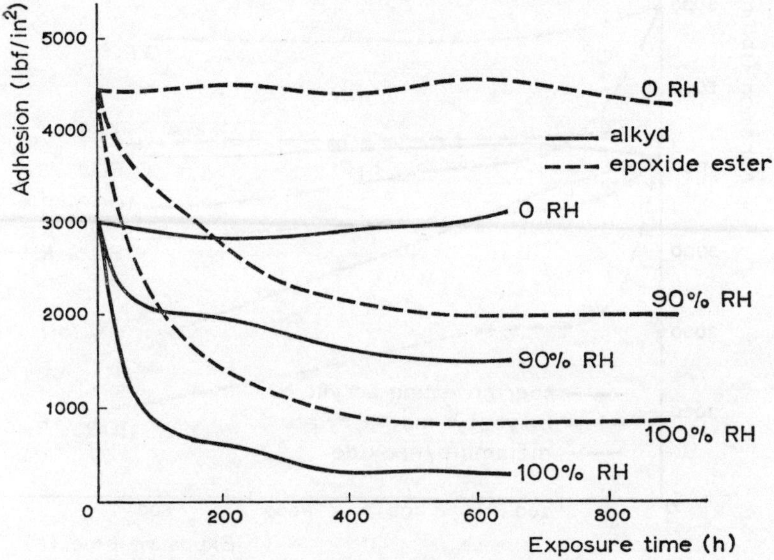

Fig. 63. Effect of humidity on an alkyd and an epoxide ester[7] (*Courtesy The Federation of Societies for Paint Technology*)

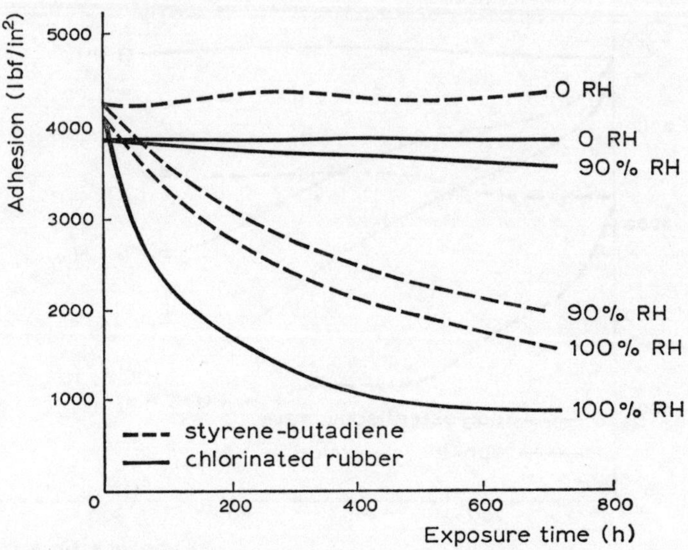

Fig. 64. Effect of humidity on adhesion of chlorinated rubber and styrene-butadiene paints[7] (*Courtesy The Federation of Societies for Paint Technology*)

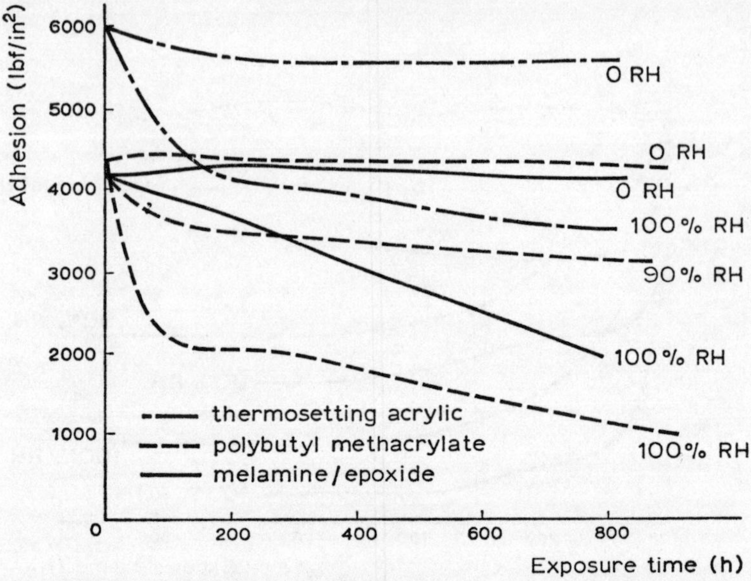

Fig. 65. Effect of humidity on adhesion of polybutyl methacrylate, thermosetting acrylic, and melamine/epoxide paints[7] (*Courtesy The Federation of Societies for Paint Technology*)

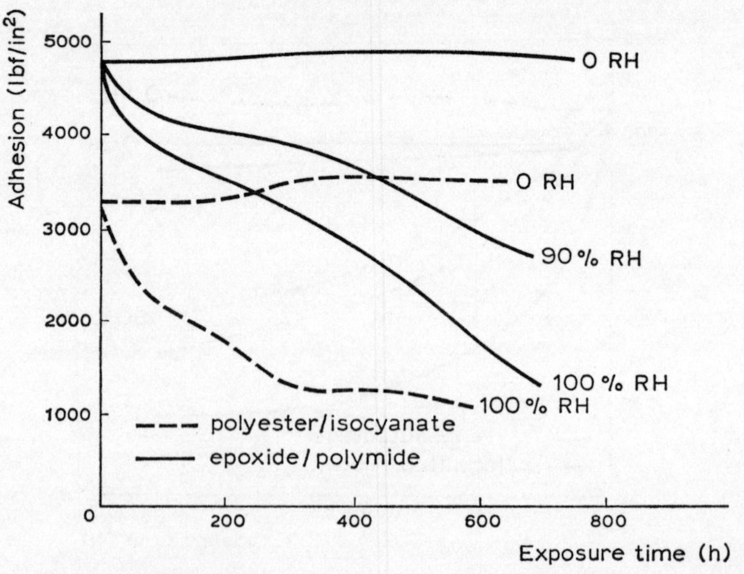

Fig. 66. Effect of humidity on adhesion of polyester/isocyanate and epoxide/polyamide paints[7] (*Courtesy The Federation of Societies for Paint Technology*)

Walker then attempted to find a correlation between the water absorption, water vapour permeability, and loss of adhesion of the various paint films. His data are shown in Table 18. Whilst there is a trend of agreement these data do not give the complete answer since, for example, severe loss of adhesion occurred with many of the paint films at exposure times much shorter than those required for significant water absorption by the detached films.

TABLE 18

Water absorption and permeability of paint films[7] *

Type of paint	Adhesion behaviour†	Water absorption (% w/w, 5d)	Water permeability (Pe × 10^{-8})
Thermosetting acrylic	Slow	1·1	1·3
Styrene-butadiene	Slow	1·6	0·9
Chlorinated rubber	Slow	1·0	1·7
(Polybutyl methacrylate)	Slow	1·3	1·2
Polyamide/epoxide	Medium	3·5	3·7
Short-oil epoxide ester	Medium	3·9	5·4
Long-oil alkyd	Fast	4·3	40·0
Melamine/epoxide	Fast	1·7	‡
Polyurethane	Very fast	5·7	33·2

**Courtesy The Federation of Societies for Paint Technology*
†The rate of loss of adhesion relative to the thermosetting acrylic
‡Film too brittle to handle.

In his discussion Walker uses a tentative explanation put forward originally by Falconer, MacDonald and Walker to explain similar adhesion loss behaviour by structural adhesives.[8]

The effect of moisture on structural adhesives

While the data on paint films was being amassed Falconer, MacDonald and Walker began examination of the response of metal adhesive joints to conditions of high humidity. It was expected that if loss of adhesion were to occur the rate would be several orders of magnitude smaller than for paint films since, it was argued, water molecules can attack over a large area of paint films but can only gain access to adhesive bonds via the periphery. Initial experiments were therefore conducted at 80°C and 100 per cent RH. Apart from this elevated temperature feature the experimental conditions were similar to those employed by Walker in his investigations. The adhesives were a heat-cured epoxide* and a heat resistant phenolic adhesive meeting the requirements of Specification DTD 5577[9] and were examined both in tension and in tensile shear on stainless steel adherends. The decay curves are shown in Fig. 67 and illustrate that the pattern of behaviour is similar to that observed with paint films. These data were the subject of a preliminary note[8] and at the time of publication an attempt was made to consider the effect in terms of Bikerman's equation:[10]

$$f_m = \frac{1}{a}\left(\frac{\zeta}{\beta} - s\right) \quad . \quad . \quad . \quad . \quad . \quad . \quad (1)$$

where f_m is the breaking stress of the bond,
ζ is the theoretical cohesion of the adhesive polymer,
a is a stress concentration factor caused by rheological differences between adjoining layers,
β the stress concentration factor, and
s the stress concentration factor due to shrinkage stresses.

Simultaneous experiments conducted at room temperature in saturated water vapour conditions produced curves of similar shape but as expected these had a much reduced rate of loss of adhesion.

Dissatisfaction with reproducibility led Kerr, MacDonald and Orman to consider refinement of the experimental techniques, and reappraisal of the phenomenon. Much of what follows concerns the original work carried out by these workers; part has been reported in full[11] while the later investigations are being prepared for publication.

In spite of Walker's success with stainless steel as substrate the adhesive investigations with this material were bedevilled by poor reproducibility of results from replicate experiments. This was particularly noticeable with tensile shear experiments. This difficulty led to a detailed examination of a range of surface etch procedures and it became obvious that none of these could be guaranteed to give standard surface properties, possibly due to slight differences in the work hardening of the surfaces during rolling of the sheet. For this reason a standard grade of aluminium-coated aluminium alloy sheet was chosen for subsequent work, BS 2L72, using miniature lap shear test specimens.

*The system used was based on 2, 2-bis-(4-(2, 3-epoxypropoxy)-phenyl)-propane, (bisphenol A diglycidyl ether), and bis-(4-aminophenyl)-methane, (DDM).

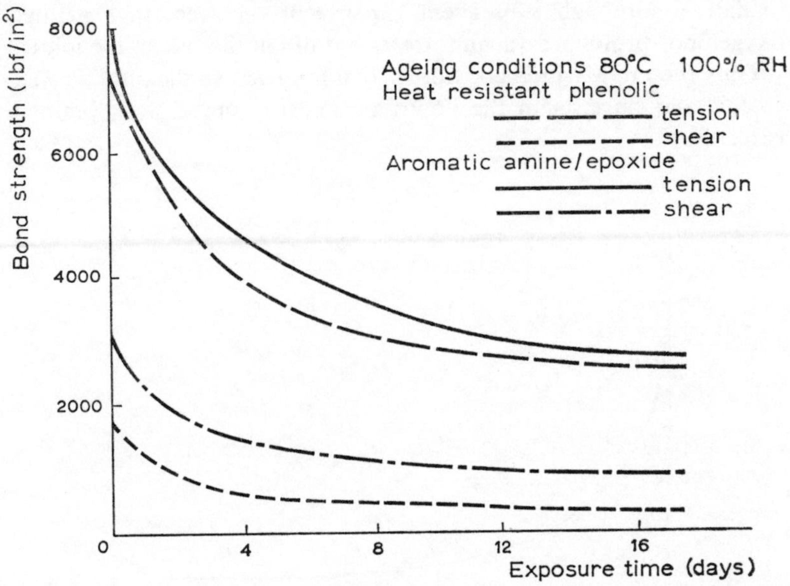

Fig. 67. Effect of humidity on adhesion of two structural adhesives to stainless steel[8]

The reasons behind the concentration on the tensile shear mode should be explained, since it is well established that this configuration leads to high local stress build-up in the loaded joint;[12] because of this, lap shear specimens are not normally used for fundamental work. Where pure shear stress is required the so-called napkin-ring test has been recommended.[13] The object of this investigation, however, was to examine *changes* in joint strength, and so it was argued that where a single adhesive/adherend combination is used the nature of the bond geometry is unimportant, provided that the geometry is not a variable. A second reason for the choice was that such joints are easy and comparatively cheap to make in large numbers. The design of the test specimen also enabled exposure to be carried out on sets of five replicates in sealed ampoules; the loaded ampoules were evacuated on a vacuum line and a known quantity of water, a slight excess over that required for saturated conditions at the test temperature, transferred into the ampoules before sealing. This feature of the experimental design allowed the effect of water to be studied in highly controlled conditions free from possible side effects such as oxidation, although subsequent experiments showed that neither oxygen nor prolonged vacuum treatment did in fact affect the joints.

Only the DDM-cured epoxide system has been studied in this way so far, and once again the now typical decay curve was obtained (Fig. 68).

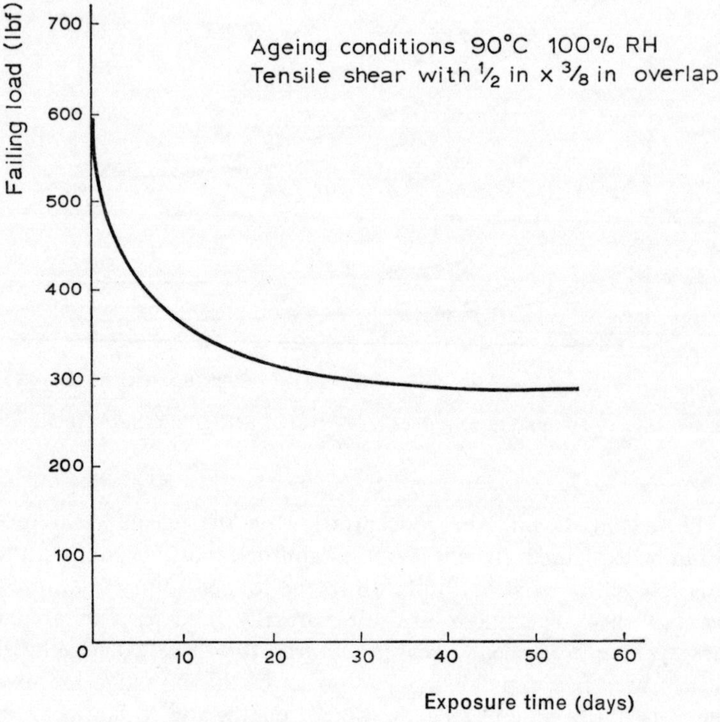

Fig. 68. Effect of humidity on adhesion of a DDM/epoxide to aluminium[11] (*Courtesy The Society of Chemical Industry*)

Discussion

The preliminary paper by Falconer, MacDonald and Walker[8] discussed the possibility of the existence of material at the interface region which is markedly more sensitive to alteration of properties by water molecules than the main bulk of the polymer. In order to test this hypothesis the tensile strength of the polymer itself was examined independently using small, circular section dumb-bells machined from castings prepared from the same resin/hardener system as was used for the adhesive experiments. These test specimens were based on a design referred to by Hounsfield.[14] The loss of strength at the equilibrium level was in this case some 10 per cent of the initial value, compared with about 50 per cent in the case of the adhesion experiments, and the time to achieve equilibrium was 1 to 2 days compared with about 30 days for the aluminium/aluminium adhesive joints,[11] or about 12 days with stainless steel/stainless steel joints.[8] In addition the time to reach equilibrium in terms of water absorbed was 2 to 3 days and the uptake at equilibrium was just over 2 per cent by weight.

Similar experiments, on adhesive joints and on polymer were conducted using a less polar hydroxylic moeity, ethanol. Marked deviations in behaviour were at once apparent. Firstly, ethanol had no effect on bond strength of the adhesive joints, but had a severe effect on the tensile strength of bulk polymer, the loss after 50 days exposure being about 35 per cent of the original, and equilibrium not quite achieved; secondly, the uptake of ethanol by bulk polymer reached 7·5 per cent by weight in the same time and equilibrium had still to be reached. The data are presented graphically in Figs. 69 and 70.

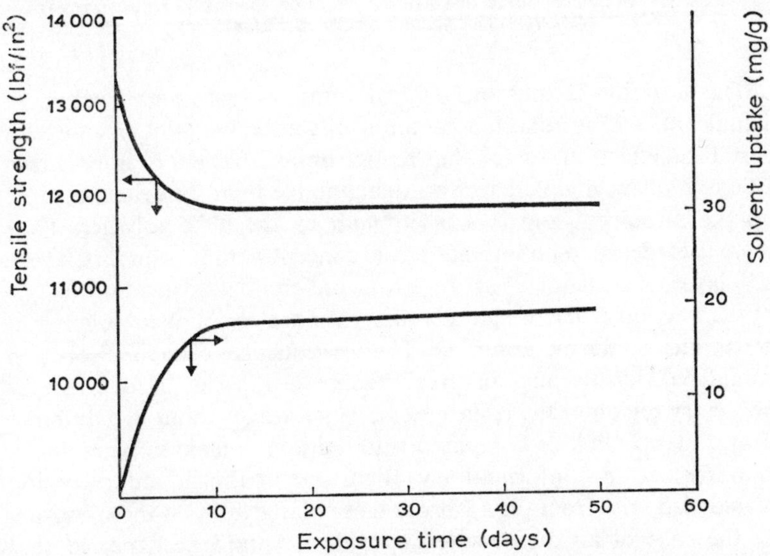

Fig. 69. Effect of water on tensile strength of a DDM/epoxide polymer at 90°C[11] (*Courtesy The Society of Chemical Industry*)

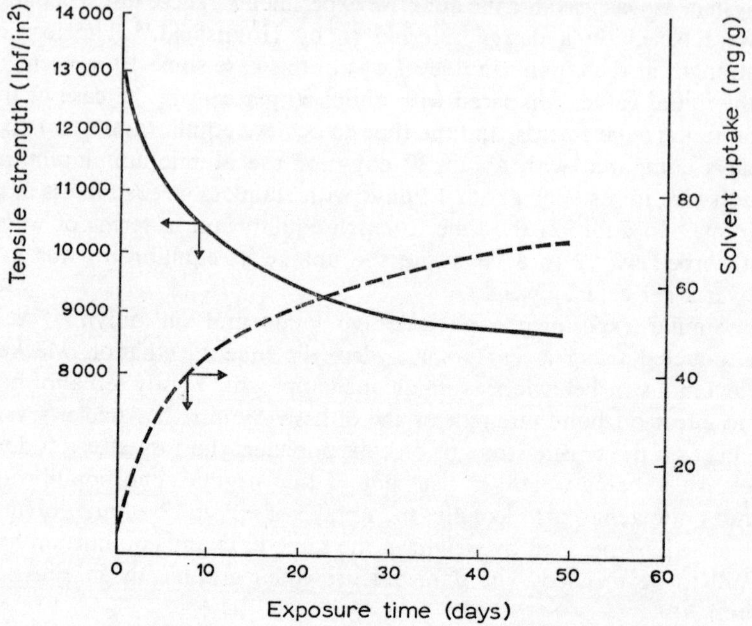

Fig. 70. Effect of ethanol on tensile strength of a DDM/epoxide polymer at 90°C[11]
(*Courtesy The Society of Chemical Industry*)

The situation is thus one of real contrast—with water vapour a significant loss of adhesion occurs in the adhesive joint but there is very little effect on, or for that matter uptake by, the bulk polymer. Ethanol has negligible effect on the adhesive joint, but there is large uptake of solvent and loss of strength by the bulk polymer. Thus there is evidence to add weight to the concept of the 'boundary layer', this layer, it is claimed, arising from preferential adsorption of one of the low molecular weight precursors of the crosslinked polymer at active sites on the metal surface. The consequences of such adsorption would be a disturbance of 'free' reactive groups in the interface or boundary region of the resin system, which would result in a different degree of crosslinking compared with that of the main polymer mass. A difference in conformational alignment in the boundary region would also arise from the adsorption of polar groups on the substrate. In the case of an amine-cured epoxide it would be expected that hydrogen bonding to substrate would constitute the principal adhesion forces; hydrogen bonding between such groups would also be responsible for much of the interchain bonding contributing to the cohesive strength of the boundary polymer. The element of order will be at a maximum at the substrate and hence disruption of hydrogen bonds in the boundary region will bring about failure of the stressed joint by cohesive rupture of the weakened boundary polymer. Ethanol is adsorbed at hydrophilic surfaces less readily than water. It is argued therefore that ethanol is more readily accepted into the bulk polymer than water, whereas in the boundary polymer region the reverse applies. A recent paper by Kwei[15] presents an additional concept which merits consideration in this context and is discussed later.

It is interesting to comment at this stage on the reported presence of 'water soluble' layers between certain paint films and metallic substrates. Bullet and Prosser[16] and Walker[7] have presented evidence of this effect but the analogy to a 'water soluble glue' is probably an over-simplification. The observations with certain paints are consistent with the concept of a boundary region polymer of different constitution. Walker comments that although he found soluble material with an alkyd and an epoxide ester paint, he was unable to detect such material with his polyurethane paint; yet this was the most sensitive to water attack (Fig. 66). If the principle of highly water sensitive polymer, as opposed to soluble low molecular weight bodies, in the boundary region is considered then Walker's findings are explained.

The mechanism proposed by Kerr, MacDonald and Orman[11] postulates the rapid adsorption of water molecules at the metal or more correctly metallic oxide surface. The rate of movement of water over such surfaces is extremely rapid and is much faster than simple diffusion through the bulk polymer. Less polar molecules are less efficient in their displacement capabilities, being less hydrophilic, and so have a greater influence on the less hydrophilic bulk polymer. A convenient way of explaining the difference in the effects of water and ethanol is to consider the values of the so-called solubility parameter, δ, for each solvent.[17] This parameter, related to the cohesive energy density, may be expressed as

$$\delta = \left(\frac{\Delta E}{V}\right)^{\frac{1}{2}} \quad \quad (2)$$

where E is the energy of vaporisation at zero pressure, and
V is the molar volume.

The larger the value of δ the greater the capacity for hydrogen bonding, or in the case of the boundary polymer, adhesive hydrogen bond disruption.

The values of δ for water and ethanol are 23·4 and 12·7 $(\text{cal/cm}^3)^{\frac{1}{2}}$ respectively. Subsequent unreported work by Kerr, MacDonald and Orman using methanol ($\delta = 14{\cdot}5$) and formamide ($\delta = 19{\cdot}2$) has shown that the disruptive effect on adhesion is consistent with that predicted from their solvent parameter values.

The arguments above have been constructed on the basis of the behaviour of one typical aromatic amine-cured epoxide resin adhesive system, and it would be unwise to regard the effect as universal, although in general terms the theory should apply. De Lollis and Montoya[18] have examined two structural adhesives of quite different character, a nylon/epoxide and a nitrile/phenolic type. Although it is difficult to obtain precise correlation between their data and those reported by Kerr, MacDonald and Orman,[11] it is nevertheless apparent that De Lollis and Montoya observed a rapid loss of strength with their epoxide adhesive in wet conditions (Table 19). The less polar nitrile/phenolic adhesive suffered only a slight loss of strength in the same conditions. This adhesive had an initial strength only half that of the nylon/epoxide and it would seem from its behaviour and constitution, therefore, that the adhesion would be largely due to dispersion forces.

There is no implication in the above statement that an adhesive which relies on dispersion forces for its adhesion may not be so sensitive to the movement of water molecules into the interface as would an adhesive which is hydrogen bonded to the substrate. It has been demonstrated that water can rapidly displace less polar molecules, such as benzene, absorbed on alumina surfaces[25] and there is no reason to suspect that water will not displace polymer with low polarity segments in the same way.

The essential differences between the adhesive systems examined by De Lollis and Montoya depend on the nylon/epoxide having lower molecular weight, more polar precursors than the nitrile/phenolic; it thus has a boundary layer structure significantly different from that of the main epoxide polymer mass, and in consequence more sensitive to the plasticising action of water, whereas the differences between the comparable regions of the nitrile rubber based adhesive are less intense. Normally adhesive bonds fail in cohesion and several authorities agree on the unlikely occurrence of true adhesion

TABLE 19

Effect of humidity and water immersion on the shear strength* of two structural adhesives[18]†

	Nylon/epoxide adhesive		*Nitrile/phenolic adhesive*	
Exposure time (month)	Humidity cycle‡	Water immersion	Humidity cycle‡	Water immersion
0	4 370	4 370	3 052	3 052
2	1 170	2 890	2 180	2 740
6	950	1 700	2 370	2 280
12	795	500	2 380	2 380
18	1 025	200	2 350	2 640
24	850	120	2 440	2 390

*Values quoted are means of five for the humidity exposure and three for the water immersion tests expressed as lbf/in^2 from standard $\frac{1}{2}$ in × 1 in lap joints.

†*Courtesy Wiley.*

‡Humidity cycle of 93 per cent RH between 149°F and 89°F with a cycle time of 48 h.

failure.[19, 20, 21] Sharpe and Schonhorn[1a] have commented on the fact that no information on the thermodynamic aspects of adhesion can be gleaned from joint fracture data—the forces to fracture are very much smaller than those theoretically required.[21] If we consider, however, that the cohesive strength of the boundary polymer is high, being as it is in a more ordered conformation than the bulk polymer, then the normal behaviour is explainable. If in addition this boundary polymer has a higher density of interchain hydrogen bonds than the random bulk polymer (due to the chemistry of the pre-gelation stage), then it may be expected to suffer much more loss of cohesive strength in humid environments; consequently the joint will fail apparently in adhesion, i.e. within the first few molecular layers of the adhesive polymer, and at a low value because of the severe disruption of the cohesive structure within that region. Where the chemistry of the polymer system does not allow the creation of a boundary region polymer having such great differences in fine structure from the remainder of the mass then it will be expected that no significant change in the behaviour of the adhesive joint will result from exposure to a humid environment. Such is the case with the nitrile/phenolic adhesive.

It is of interest at this point to discuss the behaviour of the cross-linked adhesives in terms of strict polymer network theory.

Kwei[15] has investigated two amine-cured epoxide polymers in the context of their time to break, t_b, at constant stress, σ, at a number of temperatures and in dry and 100 per cent RH conditions. He has applied Bueche's treatment[22] relating t_b to the jump frequency, ϕ. In Bueche's model the polymer chain passes through the cross-section of the sample at a number of places and each segment passing through the cross-section contributes to the support of the load. At steady load conditions the number of load-bearing segments decreases with increasing time as a result of chain relaxation, the relaxation time being related to ϕ. The equation relating stress, σ, to breaking time, t_b, is

$$\sigma = -(kT/V)\ln t_b + \psi \quad . \quad . \quad . \quad . \quad . \quad . \quad (3)$$

where

$$\psi = (kT/V)\ln(2F^2/\pi^2\sigma^2V^{4/3}) \quad . \quad . \quad . \quad . \quad (4)$$

In these equations V is the volume of a segment, k is the Boltzmann constant, and the function ψ is nearly constant.

Kwei examined the fracture characteristics of films prepared from 2,2-bis-(4-(2,3-epoxypropoxy)-phenyl)-propane (Bisphenol A di-

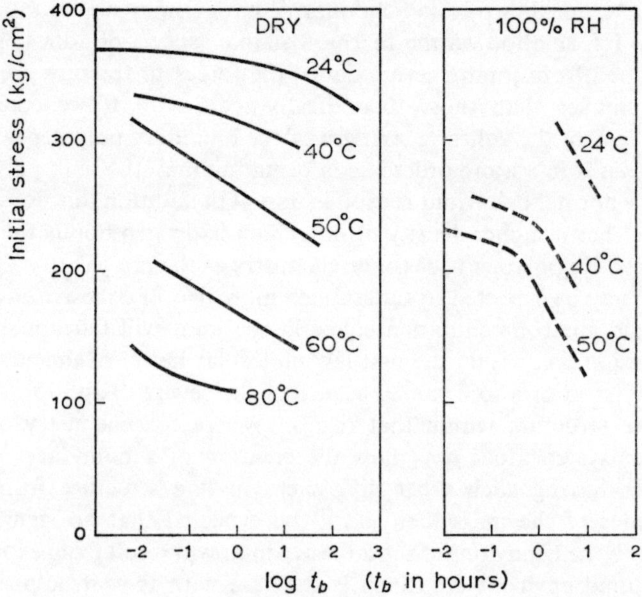

Fig. 71. (a) Basic plots

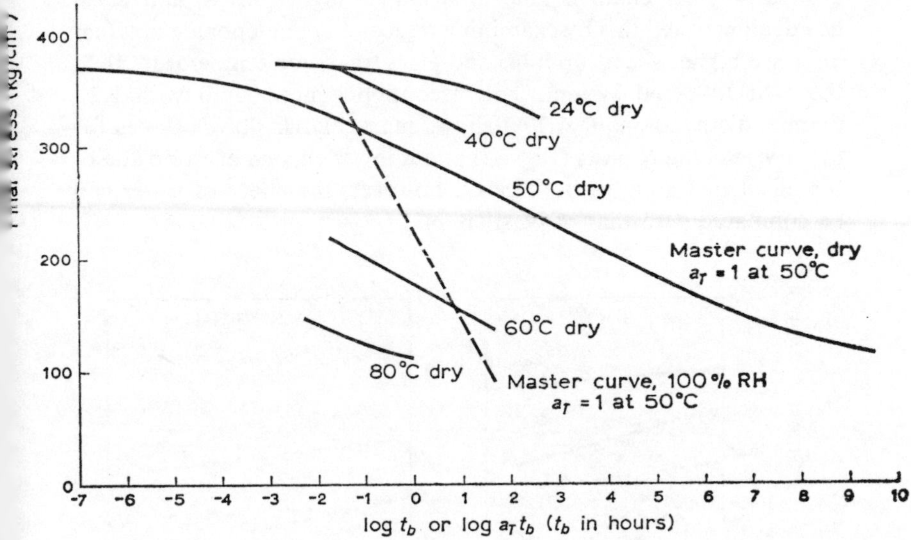

Fig. 71. (b) Master plots (*Courtesy Wiley*)

glycidyl ether) cured with two aliphatic amines. The plots of stress, σ, versus log t_b for one system, cured with hexamethylene diamine (HMDA), at a number of temperatures are shown in Fig. 71(a) for both dry and 100 per cent RH conditions. From the individual curves for each temperature condition a master curve can be constructed by introducing a shift factor, *a*. The two master curves are shown in Fig. 71(b). From a calculation of the apparent activation energy of fracture Kwei goes on to show that the jump frequency in wet conditions, ϕ_w, is many orders of magnitude larger than that for dry conditions, ϕ_d. At 50°C it is shown that

$$\log \frac{\phi_w}{\phi_d} = 8{\cdot}2$$

This increase in ϕ under wet conditions is consistent with decreases in *E* and *V* and Kwei uses this relation to propose that water molecules may interact with hydroxyl groups which are engaged in interchain bonding in the epoxide polymer.

Now a direct consequence of Equation (3) is that it predicts a large value of t_b, at a given stress, for a polymer of restricted mobility or low jump frequency ϕ. Conversely, for a given value of t_b, a less mobile polymer chain is able to sustain a higher stress, and Kwei has demonstrated this by examining a second amine/epoxide polymer having a higher shear modulus and glass transition temperature than the HMDA-cured system. This second polymer, cured with 2,2′-diaminodiethylamine (DETA) shows markedly shallower slopes for the σ versus log t_b plots (Fig. 72) and a lesser change of slope due to the effect of water. In both cases, however, the effect of water can be compared with that of plasticisers.

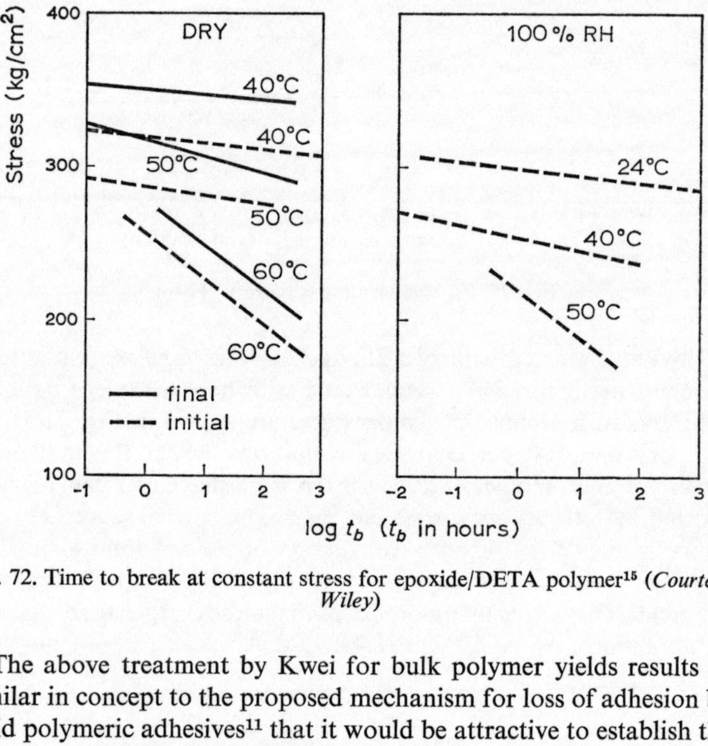

Fig. 72. Time to break at constant stress for epoxide/DETA polymer[15] (*Courtesy Wiley*)

The above treatment by Kwei for bulk polymer yields results so similar in concept to the proposed mechanism for loss of adhesion by rigid polymeric adhesives[11] that it would be attractive to establish the existence of the time-temperature relationship for adhesive bonds of the type described in this paper.

Recovery of adhesion after high humidity exposure

Walker considered the effect of allowing water-soaked paint films to dry out before determining the adhesion to substrate.[7] As with his data on loss of adhesion in wet environments it is difficult to formulate a simple statement of the effect, due no doubt to the very different chemical natures of the range of surface coating polymers involved. One common feature does appear, however; all systems show some recovery of adhesion, the alkyd being particularly noteworthy in that its final adhesion was higher than the original. In general it was found that drying over phosphorus pentoxide for three days at room temperature was not really effective, whereas simple oven-drying for four hours at a range of temperatures between 40° and 120°C resulted in more significant recovery. For the purposes of the present discussion the behaviour of the thermosetting coatings is of particular interest: neither the acrylic nor the melamine/epoxide recovered more than 60 per cent of the adhesion lost, even when heated to 120°C.

Experiments with a DDM-cured Bisphenol A diglycidyl ether adhesive have enabled Kerr, MacDonald and Orman to demonstrate that adhesion to aluminium lost on exposure to saturated water vapour environments is to some extent recoverable.[23] The recovery was effected by prolonged pumping-out of the exposure ampoules[11] prior to strength testing. It would appear that in these conditions, designed to carry any water-removing process to equilibrium, the adhesive recovers to a constant value no matter the length of exposure to the high humidity environment, this recovered value being some 85 per cent of the original bond strength. The reasons for this limitation to recovery are not as yet established. On the basis of the hydrogen bond disruption theory described by these workers[11] it can be argued that once the highly oriented boundary layer has suffered damage to its structure by ingress of water it is very unlikely that the same degree of order can be achieved after removal of the water molecules. Another view has been expressed by Eley and Rudham[24] from studies of enthalpies of immersion of aluminium oxides in a number of media. They suggest that progressive hydration of the oxide surface layer leads to an increase in thickness because of the greater molar volume of the hydrate. Subsequent degassing can cause, therefore, a reduction in volume leading to the production of a cracked surface layer. Electron micrography tends to support the existence of a fissured oxide surface, and thus gives credence to the possibility of mechanical damage at the interface being responsible for the irreversible component of adhesion loss on exposure to high concentrations of water vapour. At the same time it must be borne in mind that in normal circumstances aluminium oxide surfaces carry a number of adsorbed layers of water; workers at the University of Louisville[26] have used tritiated water to demonstrate that final removal of water from the surface only occurs at temperatures in excess of 700°C *in vacuo*. The practical realism of the hydration/dehydration cycle may not therefore be particularly great.

Conclusions

Adhesive bonds involving metallic substrates may be expected to suffer some loss of strength in humid environments because of the rapid ingress of water molecules at the interface region. The greatest disruptive effect is likely to occur with those polymer systems which rely most on hydrogen bonds for adhesive forces. Where the chemistry of the system allows, a boundary region particularly rich in polar groups can be formed prior to gelation; such a region will be highly oriented and very susceptible to interaction with water molecules. Similar effects occur with surface coatings, and there is some evidence from studies with paint films that hot-cured systems have lower sensitivity to water than cold-cured materials.

Acknowledgements

The author thanks the Director, Atomic Weapons Research Establishment for permission to publish this paper. He also thanks his colleagues in the Chemical Technology Division for helpful discussions; particular reference must be made to the contributions of Dr. S. Orman and Mr. C. Kerr who have collaborated in much of the work referred to in this paper.

References

1. *Adhesion and Adhesives*, Eds. de Bruyne, N. A. and Houwink, R., Elsevier, Amsterdam, 1951.
1a. Sharpe, L. H. and Schonhorn, H., Surface Energetics, Adhesion, and Adhesive Joints, *Contact Angle, Wettability, and Adhesion, Advances in Chemistry Series*, American Chem. Soc. Washington DC, 1964.
2. Voyutskii, S. S. (trans. Kaganoff, S.), *Autohesion and Adhesion of High Polymers*, Interscience, New York, 1963.
3. Deryaguin, B. V. and Krotova, N. A., *Doklady Akad. Nauk. SSSR*, **61**, 849, 1948.
3a. Skinner, S. V., Savage, R. L. and Rutzler, J. E., *J. Appl. Phys.*, **24**, 438, 1953.
4. Kiselev, A. V., *Disc., Faraday Soc.*, **40,** 205, 1965.
5. Dunkley, F. G. and Earp, D. P., *J. Oil Colour Chem. Assoc.*, **44,** 649, 1961.
6. James, D. M., *J. Oil Colour Chem. Assoc.*, **43,** 391, 1960.
7. Walker, P., *Official Digest*, **37,** 1561, 1965.

8. Falconer, D. J., MacDonald, N. C. and Walker, P., *Chem. and Ind.*, p. 1230, 1964.
9. Aircraft Materials Specification DTD 5577, H.M.S.O., London.
10. Bikerman, J. J., *The Science of Adhesive Joints*, Academic Press, New York, 1961.
11. Kerr, C., MacDonald, N. C. and Orman, S., *J. Appl. Chem.*, **17**, 62, 1967.
12. Mylonas, C., *Proc. Soc. Exptl. Stress Anal.*, **12**, (2), 1955.
12a. Greenwood, L., Boag, T. R. and McLaren, A.S., Stress Distribution in Lap Joints, *Adhesion: Fundamentals and Practice*, Ministry of Technology, Maclaren, London, 1969.
13. De Bruyne, N. A., The measurement of the Strength of Adhesive and Cohesive Joints, *Adhesion and Cohesion*, Ed. Weiss, P., Elsevier, New York, 1962.
13a. Bryant, R. W. and Dukes, W. A., Shear Strengths of Joints of Different Designs and of Various Adhesive Thicknesses, *Adhesion: Fundamentals and Practice*, Ministry of Technology, Maclaren, London, 1969.
14. Hounsfield, L. F., *The Hounsfield Tensometer Manual*, Fig. 6, Type 11(cast), Tensometer Ltd, Croydon.
15. Kwei, T. K., *J. Appl. Polymer Sci.*, **10**, 1647, 1966.
16. Bullett, T. R. and Prosser, J. L., Problems in Adhesion of Paints to Metals, *Adhesion: Fundamentals and Practice*, Ministry of Technology, Maclaren, London, 1969.
17. *Polymer Handbook*, Eds. Brandrup, J. and Immergut, E. H., Interscience, New York, 1966.
18. De Lollis, N. J. and Montoya, O., *J. Appl. Polymer Sci.*, **11**, 983, 1967.
19. Bowden, F. P. and Tabor, D., *Proc. Second Int. Cong. on Surface Activity*, **3**, 386, Butterworth, London, 1957.
20. Schonhorn, H., Adhesion to Low Energy Polymers, *Adhesion: Fundamentals and Practice*, Ministry of Technology, Maclaren, London, 1969.
21. Eley, D. D., *Kolloid Z.*, **197**, 129, 1964.
22. Bueche, F., *Physical Properties of Polymers*, Interscience, New York, 1962.
23. Kerr, C., MacDonald, N. C. and Orman, S. To be published.
24. Eley, D. D. and Rudham, R., The Influence of Metal Oxide Films on the Wetting of Metal Surfaces by Adhesives, *Adhesion: Fundamentals and Practice*, Ministry of Technology, Maclaren, London, 1969.
25. Wade, W. H. and Hackerman, N., *J. Phys. Chem.*, **69**, 314, 1965.
26. Long, J. S., Schwartz, M., Chiang, Y. S., Schuler, R. E. and Sturgeon, R. *Official Digest*, **35**, 11, 1963.

Bonding Systems for Plastics

D. F. GENTLE, A.R.I.C.

FORD MOTOR COMPANY LIMITED

In this paper you will notice that particular emphasis is placed on the economics of producing a bonded component. This approach stems from my work in the automotive industry but I believe the basic concept should be accepted by anybody using bonding techniques in production. To help explain our particular interest in this subject I would like to give you a few facts about adhesive usage in the automotive industry. Actual figures will obviously be based on Ford Motor Co. products.

During assembly of the various components into each vehicle, approximately 1·35 lb of adhesive is used. This does not include adhesives used by sub-contractors which at a rough estimate would double this figure. Assuming that 2·7 lb of adhesive were used per vehicle, the total usage would be well in excess of 1 million lb per year. These figures only include adhesives which are used in contact with plastics; a list of typical applications is given in Table 20. As adhesives are only a section of the available bonding systems used, it is obvious why we, and I assume other large manufacturers, consider cost as being of great importance. The only aspect of greater

TABLE 20

Typical Applications for Adhesives used for Adhering Plastics in Automotive Production

- Headlining adhesives
- Dielectric bonding agents for soft trim
- Edge folding adhesives on soft trim
- Adhesion of hard plastics to metal fixing clips
- Adhesion of acrylic medallions
- Adhesion of foils on reflectors
- Crash pad assembly
- Ventilation system assembly
- Vacuum adhering of PVC on steel
- Adhesion of vinyl roof covers
- Foam assembly in seat construction
- Flock spraying adhesives

- ☐ *KEMIKAL*& KEMIDOL*(OXIDES & HYDRATES)*
- ☐ *STABILIZATION AND GLASS HOUSE LIMES*
- ☐ *CRUSHED STONE PRODUCTS*

CALL:

lime products

UNITED STATES GYPSUM

A.T. HAYS
3098 PIEDMONT RD. N.E.
ATLANTA, GEORGIA 30305
(404) 261-3180

MAKE B-3711
SUBSTITUTE:
FR-3501 FOR PVC-7401B
K-90 FOR B-30E
DHP FOR H-29

importance is that the bond performs satisfactorily under service conditions and can be produced with available equipment at required production rate. In a car temperature requirements are higher than might be expected and are illustrated in Plate 28; these are typical tropical conditions.

There is a tendency in industry for bonding systems to be specified for particular applications because somebody else is using the same system for a similar application. In many ways this is an attractive approach on the basis of the risk having already been taken by the other person, but I consider this can be dangerous, especially when adhesives are being used. Information from other sources is often misleading because even slight differences in design and materials can completely change the technical and economical advantages of particular bonding systems.

Perhaps I should define what I mean by a bonding system for plastics. I consider this to be the materials and/or method used to adhere a plastic part to itself or to another substrate. This can include: (a) welding, (b) solvent cementing, (c) adhesive, or any combination of these.

In order to make a good choice of system it is necessary to have a fair knowledge of the various methods and materials available. It is intended to briefly cover the various methods before considering in more detail the problems associated with individual plastics.

Welding methods

Welding methods may be roughly divided into those methods in which heat is applied directly to the plastic material and those in which some form of energy is applied which acts on the plastic in such a way as to cause heating at the interface or in the plastic as a whole. The applied energy can be in many forms.

Direct heating methods

Hot plate welding—in this method the plastic to be bonded is heated by pressing the bonding surface on a hot plate which has a non-stick coating of PTFE. The two molten surfaces are then pressed together and held under pressure until the bond has cooled.

This method is normally used to join two parts of the same plastic together but can be used to join different plastics or even plastic to metal by heating one surface when the molten plastic acts as a hot melt adhesive on the other substrate. Careful control of temperature is

TABLE 21

Hot Plate Temperatures, °F	
A.B.S.	450
Acetal	500
Phenoxy	550
Polyethylene L.D.	360
H.D.	390
Polycarbonate	650
P.P.O.	650
Noryl	525
Polypropylene	400
Polystyrene	420
S.A.N.	450
Nylon 6, 6	475
PVC	450

necessary and a list of some of these temperatures is given in Table 21.

This method is normally only applied to flat bonding surfaces where flat hot plates may be used. These hot plates are best constructed of PTFE covered aluminium. A closely related method is used in film lamination where the film is passed over a heated section immediately prior to the laminating rollers (Fig. 73). Hot metal strips passing over the plastic surfaces may also be used as an alternative for both these methods.

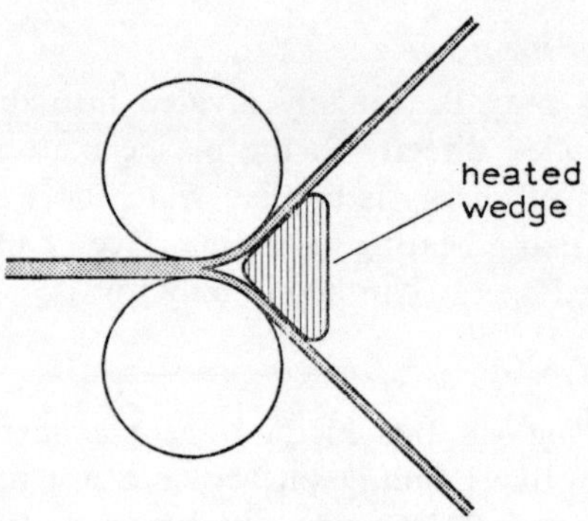

Fig. 73. (*Courtesy Wiley*)

Radiant heat welding—in this method the components are passed one each side of radiant heaters; this method can be applied to surfaces which are not flat providing good mating surfaces are used. In some cases heat shields must be included in the apparatus to protect areas which should not be heated.

Hot gas welding—is a technique used mainly for joining plastic sheet using guns of about 400–600 watt input if electric, or burning butane or propane if gas heating is used. The electric guns are usually more portable and easier to use. For easily oxidisable plastics such as polyethylene or polypropylene, either nitrogen or an inert gas must be used, but for PVC etc. dry air is acceptable. A welding rod of the same material is used to produce a fillet, the design of which is of considerable importance. A large difference between melting and decomposition temperature is necessary if this technique is to be used. The most commonly used materials are: Low density polyethylene, High density polyethylene, Polypropylene, PVC. Acrylics and polyamides have also been joined by this method.

Hot iron welding—is one of many names for the joining of two plastic items, at least one of which must be in film form, by application of heat and pressure to the external surface. In the production of polyethylene bags, an electrically heated bar is pressed against the films until the heat is conducted through the film to the interface. For thin films it is preferable to use *Impulse heating* in which a short electric pulse is passed through a resistance wire within the metal bar, the controlled heat produced being transferred to the film. Both of these methods require careful control and exist in many different forms including hot rollers. A variation of this technique is the use of a high melting film as the outer layer; this avoids the problem of cut-through at the edge of the bonds. An example of this is the bonding of sealing films to the back of acrylic reflectors in which the film is PVC polyester laminate.

Resistance wire welding—a resistance wire is placed in the joint and while the faces are pressed together a current is passed which heats the wire and melts the plastic surfaces. Normally, this would only be used where plastic is joined to itself. Obviously, metal substrates are not recommended.

Laser Beam—the high energy pulsed laser beam can be used for high speed welding especially where a sealing and cutting operation on film is required.

Indirect Heating Methods

Dielectric bonding—this technique is most widely used for bonding plasticised PVC to itself or to other compatible surfaces, but any material with a sufficiently high dielectric loss factor can be welded in this way.

An alternating electric field, from a radio frequency transmitter operating at about 27 megacycle/sec applied between metal electrodes which also act as pressure points, causes rapid reorientation of polar molecules; the subsequent energy loss is in the form of heat. The heat is generated evenly through the material but the cooling effect of the electrodes means that the hottest section is in the centre of the joint where it is most required.

Variables are: (a) Frequency; (b) Power applied per area of electrode; (c) Pressure; (d) Time; (e) Dielectric loss factor.

An increase in any one of these will give more heat and a careful balance is needed. Usually the frequency is constant.

Dielectric loss = Dielectric constant × Dissipation Factor.

and must equal = at least 0·01 and preferably 0·05.

A cooling time under pressure (dwell time) is also important and is usually about 5 sec. Olefins and styrenes are only bondable by transference of heat from high dielectric loss materials. Water acts as a great source of heat and can cause problems due to 'burning' on the bond lines.

Induction heating—an induction coil of the type used in metal heating is used in combination with iron or nickel where heat is produced by eddy currents and hysteresis. The metal can be in the form of a grid or as metal powder mixed into plastic which may be pre-moulded into gasket form. Useful where small complex shapes are involved but should not be used where strong corrosive agents could come into contact with metal.

Ultrasonic welding—a transducer acting through a resonating horn hammers the plastic surfaces together and this combined with transverse slipping causes very fast heat build up and diffusion at the interface occurs.

This is very fast and surfaces need not be very clean. Rigid plastics with high modulus of elasticity are best. Horn should be placed perpendicular to the centre of the joint surfaces.

Limitations in size are now being overcome by multiple units. The joint should be designed preferably so that flash forms internally by using non-matching tapers which leave small gaps to be filled with flash. It is also desirable that limited contact area is initially in contact (see Fig. 74). Although joints of the type as shown in Fig. 75 are often recommended it must be remembered that some difficulty could be

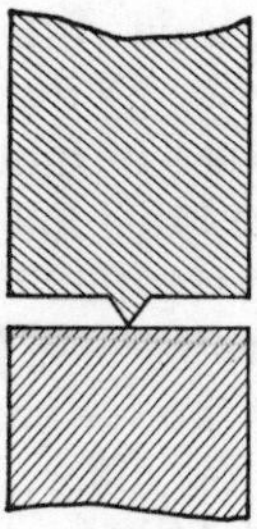

Fig. 74.

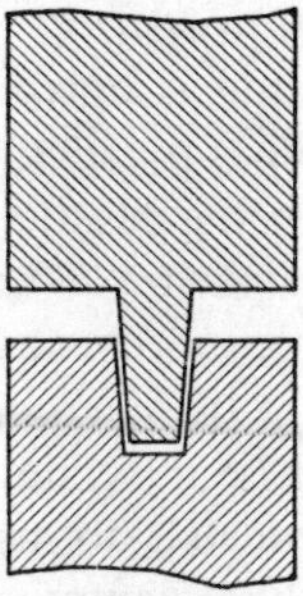

Fig. 75.

experienced in moulding this type of joint to fit where a component is greater than a few inches in diameter.

Plastics can be welded through a vacuum metallised layer which is very useful for lamp assemblies. The joint can also be made optically perfect, but, foaming can occur in the joint and any fault such as this can show in other areas due to internal reflections. Metal inserts can be forced into place by this technique. Used for bonding plastics to themselves only, except for metal inserts.

Spin welding—utilises the frictional heat produced when two surfaces of the same plastic are rotated together. Average linear velocity of joint is normally about 20 ft/sec, which is difficult for samples of less than 1 in diameter, so greater pressure or time is required. About 200 lb/in^2 pressure is normally used. Once melting occurs the rotation must be stopped. Various joint designs to reduce external flash can be used. It is usual to taper the surfaces so that initially the more central areas are heated first; this overcomes the problems of lower linear velocities at the centre. Sufficient pressure to force gases out of the joint is required. Normally only used for bonding plastic to itself for effective results. Production time is a few seconds.

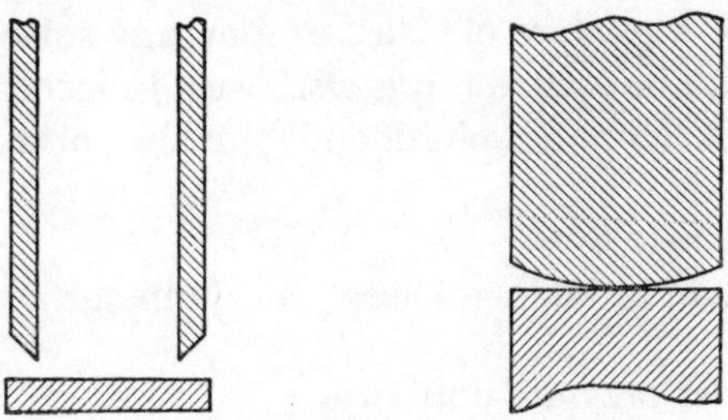

Fig. 76.

Solvent welding

Solvent welding is only really effective when used with amorphous (non-crystalline) plastics, the reason obviously being that the solvent must dissolve the plastic and enable intermixing of molecules from each piece of plastic so as to give an indefinite bond line.

It is possible to solvent weld different plastics together but this relies on the solvent being compatible with both plastics and also that the plastics must be miscible both in and without the solvent. It is comparatively unusual to use solvent welding for different materials because of these compatability problems. Solubility parameters are a guide to compatibilities of polymers.

It is normally better to use an adhesive where different materials are to be bonded, unless one substrate is porous and thus allows mechanical locking.

Problems arise with the use of solvents with rigid thermoplastics from stress cracking. The various causes of stress cracking are:

(a) Highly stressed parts in which the layer softened by the solvent is not sufficiently softened to be able to flow but is too weak to withstand internal strain.

(b) Solvents which swell plastic without causing the surface to become mobile (to flow) are absorbed by the plastic; if solvent is removed from the surface the top layer of solvent evaporates faster than its place can be taken by the further solvent migrating from the inner layers. This leaves a swollen layer under a shrinking top layer which is therefore under tensile strain. The more rigid the plastic the more likely it is that surface crazing will occur.

Both types of cracking can be partially healed by excess solvent, or by residues of higher boiling solvents, which causes the surface to become liquid and to flow. This can cause the cracks to heal. It is preferable to choose solvents which do not cause cracking but give good fast bonding.

It is usual for a mixture of a fast evaporating solvent with a high boiling solvent to be used and it is also usual to incorporate some of the resin to thicken up the solvent and to reduce shrinkage and hence internal stresses.

The bond is made by:

(a) Using medium boiling solvents on both surfaces and leaving until tacky

(b) Hold under pressure until dry

(c) Use as thin a coat of solvent as is possible

TABLE 22

Welding Methods

Plastic	Heat	Dielectric	Ultrasonic	Spin	Solvent
A.B.S.	√		√	√	√
Acrylic			√		√
Acetal	√		√	√	
Chlorinated polyether		√			
Cellulosics	√				√
Fluorocarbons	√				
Ionomers	√				
Urethane	√	√			√
Paralene	√				
Nylon	√	√	√	√	
Polybutylene					√
Phenoxy	√		√	√	√
Polyethylene L.D.	√			√	
Polyethylene H.D.	√			√	
Polypropylene	√		√	√	
Chlorinated polyethylene	√	√		√	
E.V.A.	√				
Polycarbonate	√		√	√	√
Polyester	√		√	√	
P.P.O.	√		√	√	√
Polysulphone	√		√	√	√
Polystyrene	√		√	√	√
Polyvinyl alcohol	√				√
Polypropylene oxide	√				√
PVC	√	√			√

(d) Use low stress parts, annealed if necessary

(e) Use high pressure.

Actual solvent mixtures will be given under the individual plastics. It is preferable to carry out bonding in a warm dry situation to avoid condensation due to solvent evaporation cooling the part. Heating the joint during bonding is not recommended as this can cause stress cracking as the solvent leaves the surface layer; it can also cause bubbling in the solvent layer.

The effect of cracked surface on a plastic is to reduce the impact and tensile strength very considerably by concentrating stress at the crack. It is unfortunate that rigid, brittle plastics such as polystyrene and polmethylmethacrylate tend to become cracked and are also the lower impact types which become very weak when cracked.

Monomer welding

An alternative to using a solvent for welding is to use the monomer form of the plastic which may contain dissolved polymer and usually contains catalyst. The catalysts which are often peroxides can be activated by u.v. light or heat. The advantages of this method are less shrinkage in the joint and good gap filling. Stress cracking can still occur but is less likely.

Adhesive bonding

With the types of bonding systems considered so far the plastic is bonded to itself or to a very similar plastic by intermixing of molecules at the interface under the influence of heat or solvent. Adhesive bonding is more dependent upon specific adhesion although with solvent based adhesives a certain amount of molecular mixing can occur.

The great advantage of adhesive bonding is the ease with which plastics can be joined to a wide variety of other substrates. The main disadvantage is that the bond is not homogenous and therefore stress tends to be concentrated in the bond area. It is most important that with adhesive bonding the joint should be designed to avoid peel forces on the bond. The comparative flexibility of most plastics makes this difficult but certain precautions can improve results.

Figure 77 shows a flanged joint of the type commonly used with plastic and it can be seen how peel forces are avoided by ensuring that the complete bond area is filled with adhesive and a bead is allowed to exude at the edges. When adhering film to a rigid base the forces can be converted to shear by adhering round an edge (see Fig. 78).

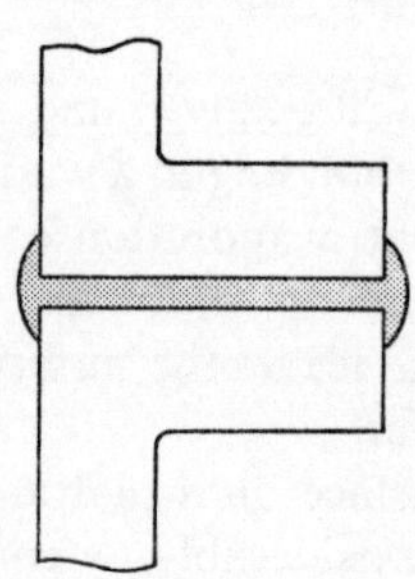

Fig. 77.

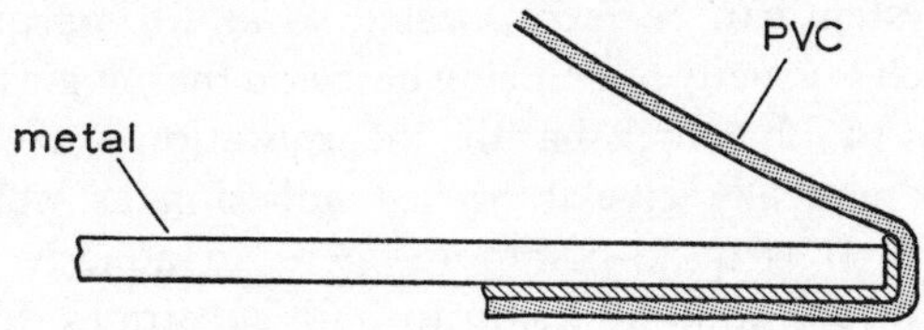

Fig. 78.

The requirements of the bond are as given earlier but I would like to emphasise the aspect of designing to do a job but not to give the strongest joint. Over-engineering is a very common fault with adhesives which does not occur with other methods because it is generally as easy to produce a good bond as a bad one with heat bonding.

Normally, if it is possible to use a heat or solvent bonding process for producing a bond, it is quicker and easier to do so if the equipment is available.

Any combination of materials can be bonded using an adhesive but with some materials the strength of the bond is quite low. For example, it is fairly easy to bond polythene to most surfaces using a permanent tack adhesive, but the ultimate bond strength is low and is subject to cold flow. If, however, the polythene surface is oxidised with chromic acid quite good bond strengths may be produced using rubber based adhesives. Surface pre-treatments will be considered under the individual plastic.

Adhesives may be used in many different forms and the various types can be divided into general groups:

(a) Solvent based, (b) Latex, (c) Film, (d) Resin with curing agent, (e) Moisture curing, (f) Hot melt, (g) Permanent tack, (h) Monomeric.

Each of these groups have certain advantages and within these groups there are several alternative materials. I intend to cover these groups to briefly try to show the advantages and disadvantages of the various materials for bonding plastics.

(a) *Solvent based*—these adhesives are normally rubber plus resin dissolved in a mixture of solvent. They are the only group of adhesives where a solution of the substrate can be formed which intermixes with the rubber in the adhesive.

Modifying resins are normally incorporated to improve tack and to modify hardness and heat resistance. Stabilisers are necessary where heat resistance or exposure to u.v. light are required.

A curing system can be incorporated so as to improve the heat resistance which is usually poor, being based on thermoplastic rubbers. It is important to remember that the incorporation of a curing agent does not automatically give improved adhesion as well as better heat resistance. It is quite common where an adhesive such as a neoprene gives very good adhesion to both substrates, for example, PVC to acrylic paint, that a curing agent such as isocyanate will cause complete adhesion failure to one or both surfaces. The same effect may be found by over-cross-linking of metal to metal adhesives. If the substrate is soluble in the adhesive this effect does not occur, and considerable increase in bond strength and heat resistance can be obtained by adding a curing agent.

The basic rubbers in common use are:

(1) Natural. Not widely used due to poor ageing, heat and environmental resistance. Reclaim rubbers are used in non-critical areas.

(2) S.B.R. Good tack range and grab, with low cost, but temperature resistance is low and it is affected by plasticisers. It is difficult to obtain with high stain resistance for PVC.

(3) P.I.B. Used mainly for permanent tack formulations.

(4) Neoprenes. The most widely used synthetic rubber and has a good tack range which is variable by changing the rate of crystallisation. Solvents normally contain toluene plus M.E.K. or esters, but hexane or any other aliphatic content can be incorporated. Isocyanate curing agents can be used for curing.

Adhesion to most surfaces is better with neoprenes than with other rubber based resins. Neoprenes are liable to stain and to soften under the influence of plasticiser.

Heat resistance and tack depend on resin used which also determines dermatitic effect (p-tert-butyl phenolic resins are not recommended). 250°F is the best softening point available without curing. Staining is due to two factors:

1. Impurities in the adhesive, which migrate to the surface of the plasticised material due to a form of solid-liquid chromatography, are decomposed to dark products under influence of u.v. light.

Greater resistance to plasticiser attack on the adhesive layer tends to reduce this effect and can be overcome by purification of adhesive components. Use of a curing agent can also reduce this effect, due to increased resistance to plasticiser and reaction of isocyanate with impurities 'locking' these impurities in adhesive layer.

PLATE 28
See page 143

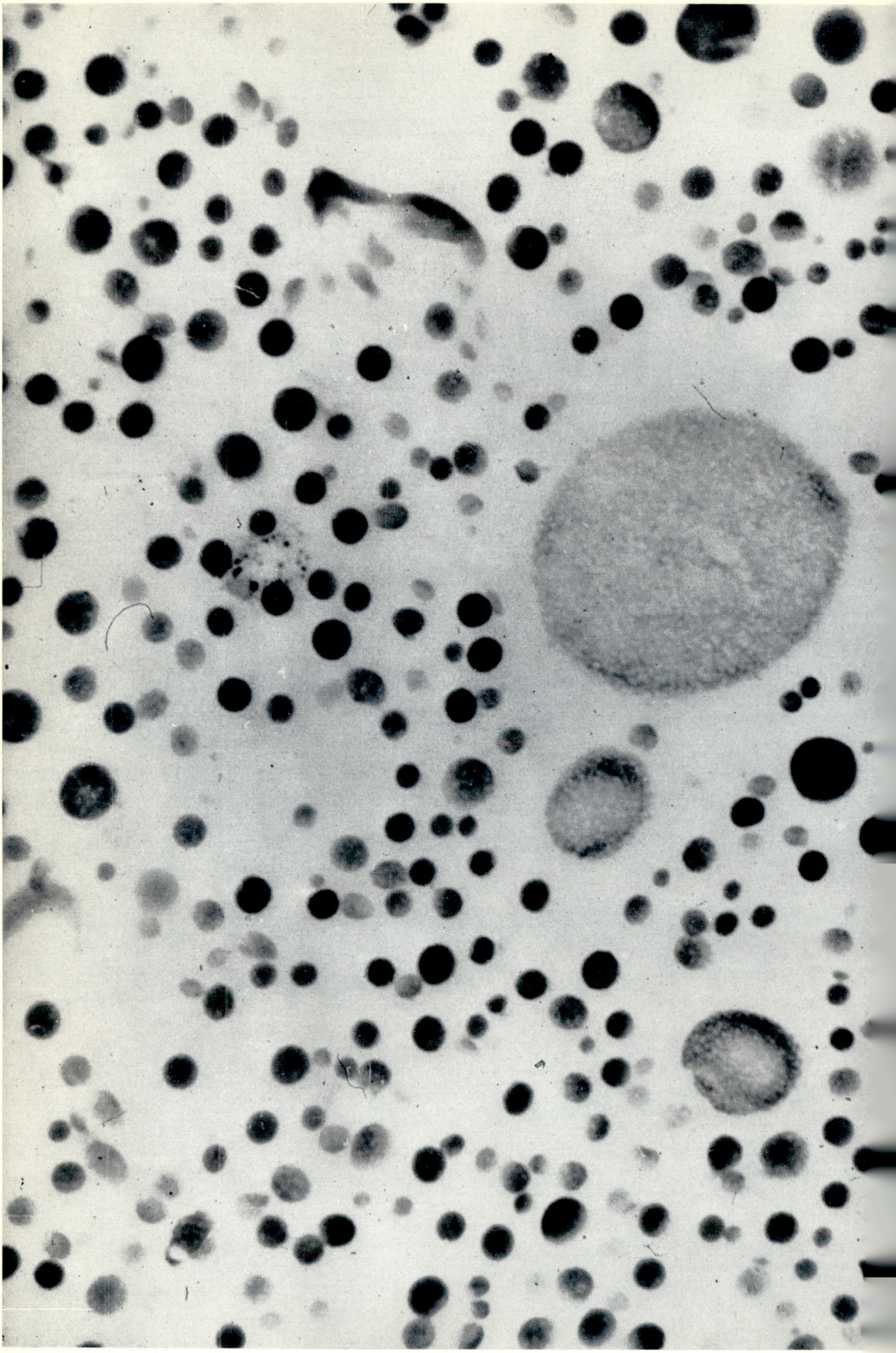

PLATE 29 Section of ABS, polybutadiene darkened with osmium tetroxide. Large area approx 1μ (*Courtesy BXL Plastics*)

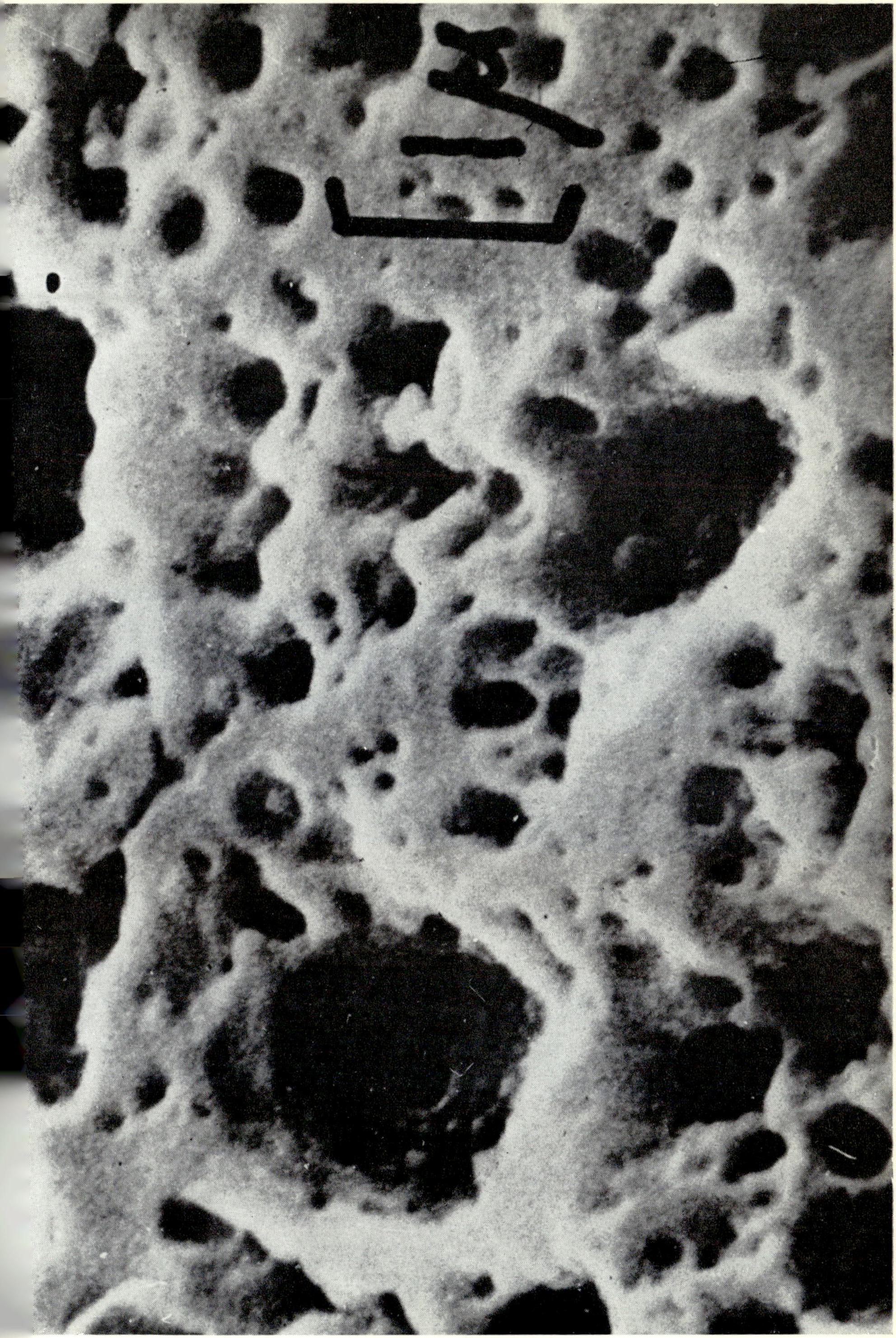

PLATE 30 Etched ABS surface (*Courtesy BXL Plcstics*)

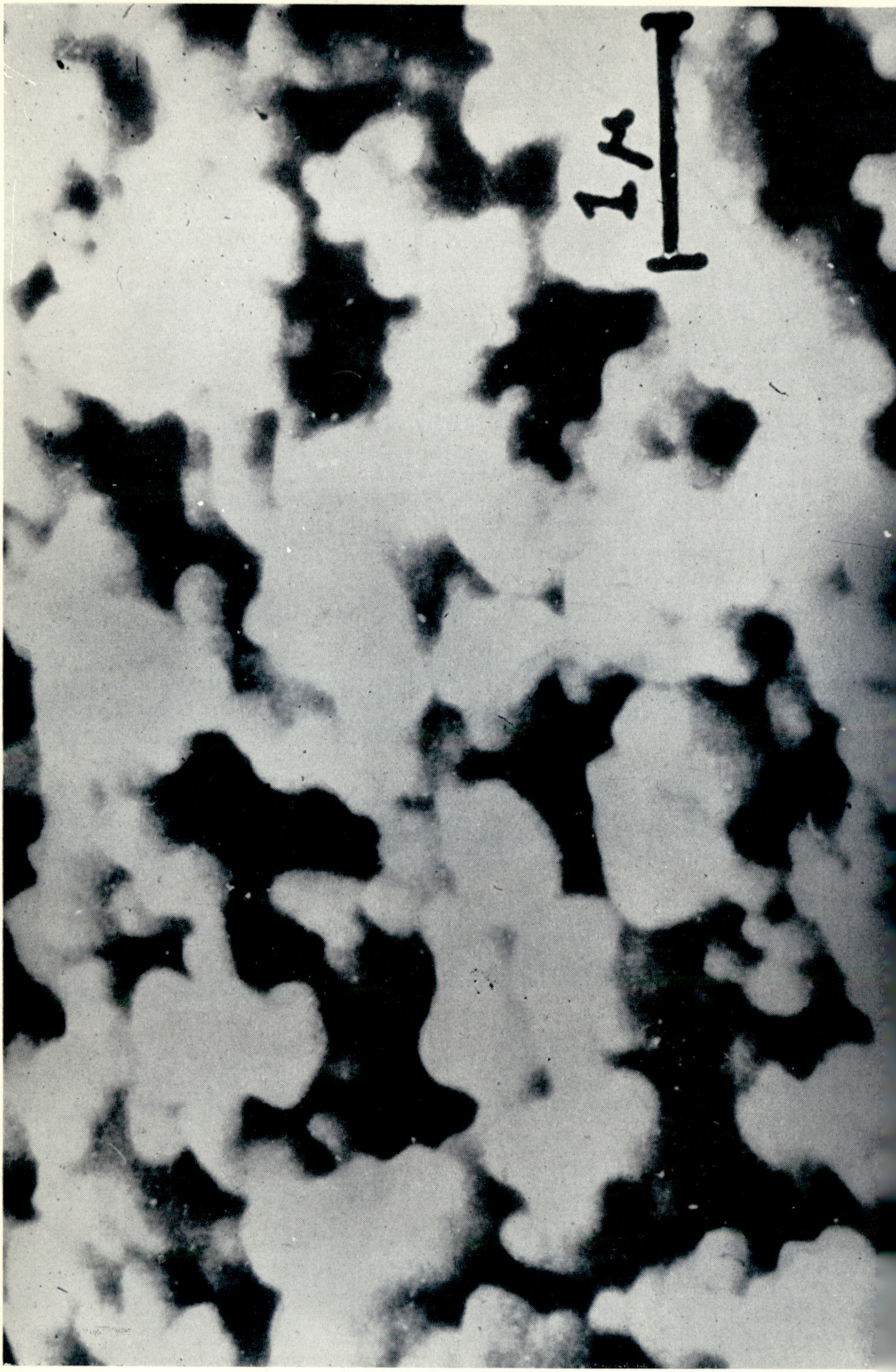

PLATE 31 Underside view of copper plated on to etched ABS and ABS removed by burning (*Courtesy BXL Plastics*)

2. Thermal decomposition of neoprene gives HCl which causes autocatalytic decomposition of PVC and eventual chromophor formation. Stabilisers in the PVC and adhesive help but these are used up by HCl. With a poorly stabilised PVC a purple stain can be caused by almost any adhesive due to the removal of stabilising plasticiser by solvent or by transference to adhesive layer. A stain caused by adhesive is yellow/brown on a well stabilised PVC.

(5) Nitriles. Normally compounded with phenolic resins which may be of the curing or non-curing type. Solvents are ketones and/or esters. Very useful for bonding plasticised materials because of resistance to plasticiser attack. One of the easiest rubber bases to make non-staining and can also be non-yellowing. Maximum heat resistance is usually about 200°F which can be increased considerably by cross-linking with isocyanates.

Use of self curing systems with curing phenolics can only be effective above 280°F which limits the use to thermosetting plastics and higher softening thermoplastics; some of these can be excellent for bonding to metals, even in presence of oil.

(6) Polyurethanes. Normally based on polyester resins which are reacted with isocyanates. Pre-reaction to give a polyurethane/polyester resin with hydroxyl end groups is normally carried out. Polyurethane adhesives can be used without curing but heat resistance and solvent resistance is greatly improved by using a diisocyanate or triisocyanate curing agent. An excess of isocyanate above the theoretical quantity required for cross-linking is always used. This reacts with any water present and helps to give a chemical bond to many surfaces. Polyurethanes give greater scope for formulation of the base polymer than other rubbers and any bond hardness can be produced without the addition of modifying resins. By mixing low and high molecular weight resins a high degree of tack can be produced. Although they tend to be expensive some of the advantages make polyurethanes very attractive.

Advantages:

Non-staining.

Range of formulation.

Cross-linking of resins to give a pure rubber without thermoplastic resin contact.

Pot life can be several days in solvent.

Formulation of some types will give a permanently tacky adhesive film which then cross-links to give a very high heat resistance. This allows a wide assembly time and pressure is not needed on the joint.

Inhibited curing agents can be incorporated to give an adhesive which does not have any pot life problem but will cure after heating. This is normally carried out during a reactivating process on the previously dried adhesive film.

Disadvantages:

Suspect hydrolysis resistance.

Yellowing with light.

Reaction with excessive water to give foam.

(b) *Latex based adhesives*—Latex based adhesives are not often used for bonding plastics, except where the latex is purely a method of applying a film of adhesive which is then heat reactivated, or is permanently tacky when the water carried is removed.

Many types of material can be applied from latex and it is widely used where films are to be coated or where a porous substrate is being used. Examples are in the preparation of tapes, the laminating of thin films where heat is used to reactivate the adhesive, and in the use of PVC copolymer latexes as bonding agents where flexible PVC sheeting is to be dielectrically welded to a porous substrate such as hardboard.

(c) *Film Adhesives*—Film adhesives are simply the residues of solvent or latex based adhesives which have been either cast on release paper (free film) or coated on to one of the substrates.

The free film adhesives can be any thermoplastic resin and may contain catalysts or curing agents if these only operate at high temperature. The film can be cast so as to include non-woven or woven fabrics to add strength. The advantages are that film may be blanked out as a gasket to give a very even and controlled film thickness.

Reactivation can be by either solvent or heat; with most thermoplastics solvents are normally used but with thermosets and a few of the higher melting thermoplastics, heat reactivation may be used. Heat reactivation is a clean fast method and where one substrate is a metal, can be very useful. This principle is widely used in vacuum covering where the hot plastic sheet is drawn onto a heat reactivatable adhesive previously coated onto the object to be covered. The requirements are very severe and for this type of part we have found that a heat reactivatable polyurethane containing an inhibited curing agent gives very good results. Production rate is very high.

(d) *Hot melts*—Hot melt adhesives originated before normal adhesive technology was even dreamed of, and of recent years they have made a dramatic recovery with the introduction of new resins.

Although most thermoplastics can be used as hot melts in either solid or film form, (e.g. polythene, polyester, PVA, PVC, nylon) most work and progress has been made on ethylene/vinyl acetate copolymers (normally blended with waxes) and polyamides.

Application can be made by hot roller, melted strip or in bulk from a nozzle. Films of hot melts are used, as discussed under film adhesives, except that solvents are not usually effective for re-activation.

The advantages of hot melts are speed and lack of solvents; this allows a thick joint to be made with little shrinkage. The disadvantages are that application temperature is very high whereas maximum useable temperature is quite low. For EVA type the maximum temperature at which a 1 lb weight could be supported in shear is about 80°C. It is claimed that this can now be overcome. Most of the high temperature EVA formulations tend to have a high melt viscosity. Polyamides have a much sharper melting point but tend to be more expensive which has limited their use.

Hot melts are often rejected as bonding systems for thermoplastics because of the high application temperature; but in fact even very highly stressed mouldings can be bonded, using a large quantity of adhesive, without distortion or even affecting metallising. Because of their high speed of application and their relatively low price I think there is a great future in hot melts for bonding plastics.

TABLE 23

Hot Melt (EVA) Shear

Plastic	Shear strength (lb/in^2)
Acrylic	350
Polypropylene	140
Acetal	120
A.B.S.	360
Carbonate	340
Alkyd	250
Steel	300
Steel (oiled)	100 (poor peel)

The low heat resistance of hot melts can be overcome with the polyamide types by incorporation of epoxy resins which cross link the adhesive after heating. This does limit the types of application equipment used.

When used for bonding to metals the chilling at the surface of the metal must be allowed for. Typical bond strengths obtained with EVA type of adhesives are listed in Table 23.

(e) *Resin with curing agent*—Most adhesives of this type give very strong cohesive layers and when failure occurs with a plastic substrate it occurs at the interface. The excellent adhesion of many of this class of adhesive to metal makes them very useful where plastics have to be jointed to metal, but, the adhesion to many thermoplastics can be extremely poor in peel.

The main types are: epoxy, polyester, phenol/formaldehyde and resorcinol/formaldehyde. The main disadvantages of all of these resins is the long cure time which normally does not lend itself to high production rates, but, the extremely high bond strength sometimes obtainable make these materials essential. When using curing adhesives it is always best to treat the surface of the plastic whether thermoplastic or thermoset. I would suggest that the following procedure is used.

(a) *Thermosets*

Remove release agents with water or solvent depending on type of release agent used. The use of soaps or PVA is recommended for release agents and silicones should never be used in the same factory if possible.

Abrade surface to be bonded.

Wipe with solvent such as I.P.A.

Dry with warm air.

Make bond by coating both surfaces and pressing together until cured; it is most important to avoid voids.

Heat bond if possible during cure to give best results.

(b) *Thermoplastics*

Anneal, wipe with solvent which does not affect the plastic.

Etch with chemical known to attack surface groups of the plastic (i.e. chromic acid for polyethylene, A.B.S. etc. Na in liquid NH_3 for PTFE etc.).

(This has the effect of giving a key or produces a more reactive surface; recently it has been suggested that a stronger surface layer is produced. For A.B.S. it has been shown that during plating processes the key is purely mechanical and a similar effect might be expected with adhesives without solvent (see Plates 29–31)).

Dry.

Bond at room temperature so as to reduce stresses in the joint.

(1) Epoxys—come in many types; the most useful for work with thermoplastics I consider to be bisphenol A/polyamide types because of the good flexibility over a wide range of temperatures. For use with thermosetting plastics it is usually best that the adhesive should match the resin in the laminate (including fillers) where the addition of these do not affect the wetting of the adhesive. Although not specifically designed to be used with plastics the addition of silane coupling agents or primers is often desirable. Epoxide resins do not bond well to plasticised plastics.

Very good results have been reported in leading publications dealing with bonding of untreated thermoplastics using epoxy adhesives, but,

(i) Peel strength is usually so low the bond is useless, for example with A.B.S.

(ii) Where abrading of surface is recommended, it should be remembered that this could affect the impact resistance of the parts, and that in general any scratching of thermoplastics should be avoided, especially on rigid, brittle plastics, unless it can be confined to the bond area.

One system which does seem to be successful is the addition of a small amount of solvent to the adhesive. This solvent should be chosen to dissolve the plastic and will obviously only work on non-crystalline plastics. The surfaces are coated and the solvent allowed to evaporate. The bond is then made in the normal way. Good metal to A.B.S. bonds have been made in this way where solventless bonds have failed.

(2) Polyesters—unsaturated polyesters containing styrene and cured with organic peroxides are not normally successful for joining thermoplastics except where the free styrene acts as a partial solvent.

Bonding of glass reinforced polyester to itself or to bare metal is very good with this type of adhesive. If possible the bond should be made before the polyester resin in the laminate has fully cured. It can also be a very useful adhesive for other reinforced thermosets; one of the most surprising properties being its ability to adhere to asbestos reinforced phenolic, containing stearates, which defies nearly all other adhesives including some phenolics and resorcinols.

(3) Phenolics—Not widely used except for bonding etched fluorinated plastics, epoxy laminates with very high heat distortion and phenolics. High temperature curing is chief disadvantage.

(4) Resorcinol-formaldehyde—Colour can be a serious problem. Bonds well to many thermoplastics but abrading of surface is often required. If nothing else will do the job it is a useful material to try. A very useful material for adhering non-plasticised plastics to wood and some metals but the colour does tend to permeate some thermoplastics. Resorcinol/phenolic mixtures are often used.

(5) Polyurethanes—Solventless polyurethanes are not often used except as sealants. Much better solvent based materials exist whose properties such as pot life are much better.

In general the solventless/curing agent type of adhesives are of most use where thermoset plastics are being jointed or where time for pretreatments and long cure are available. Environments containing solvents can make these adhesives necessary.

(f) *Moisture curing*—These are mainly silicones or polysulphides but polyurethanes are available in solvent. Silicones are most widely used. Moisture curing is convenient for use in electronics etc. where new versions which do not give off acetic acid can be used. They are expensive but are useful if sealing is required.

(g) *Monomeric*

(1) Methyl cyano acrylates (Eastman 910 etc.) adheres almost any clean surface to any other clean surface to give a very good bond: They cost about 5 gns per oz. Because of the small usage required it is not always as expensive as might be considered especially if production time would be long. Again useful where small items are being handled, but on a general production line the toxicity and tendency to stick people together is a severe disadvantage. Mislaid or dropped ampoules can also be expensive.

The very thin bond line can give tremendous stresses in the joint when a plastic is joined to metal. Always try it first and age the test pieces, because some unexpected failures can occur with high temperature, humidity and/or u.v. light.

(2) Torque adhesives—these are a form of acrylic which polymerises in the presence of metals and in the absence of oxygen. Although these are normally used for sticking bolts into metal they can give very good bonds to many types of plastic. They are still being investigated but have already proved good adhesion to polystyrene, phenolic and A.B.S. Special primers which contain a metal

salt (ferric chloride for example) must be used in the absence of some metals. They are particularly useful on a production line for attaching small plastic parts such as knobs to metal rods or threads. The cure rate is quite fast at room temperature. A variety of strengths may be obtained. They are expensive, but may be applied in a very thin layer to one substrate, and will not cure until the surfaces are placed together to exclude air. Surfaces must be a fairly good fit.

(h) *Permanent Tack*—I prefer to call these adhesives permanent tack rather than pressure sensitive because in general no curing takes place when pressure is applied.

The main use for permanent tack adhesives is for adhering plastic films or in the well-known brands of tapes available on the retail market. For this use it is important that the supporting film is treated by etching or priming to give good adhesion between the film and the adhesive.

When there is a necessity to remove the tape without leaving a sticky mess on the other substrate the adhesive layer should be partially cross-linked to give better cohesive strength. Nothing is worse to handle than a tape where the adhesive comes off the film while unrolling or leaves a sticky deposit on the facing back of the film.

Films are normally: (a) PVC with about 3 per cent plasticiser, which can be biaxially orientated, (b) PVC with up to 30 per cent plasticiser, (c) Polyester, (d) Polyethelene, (e) Cellusosic, (f) PTFE.

Adhesives

(a) Natural rubber with tackifying resin, (b) Polyisobutylene based, (c) Vinyl ether, (d) Acrylic, (e) Silicone.

These adhesives may also be applied from solution or latex for adhering films or parts in production and each has particular advantages and disadvantages.

(a) Natural rubber/resin adhesives have exellent tack and can have very good water resistance but they can age to non-tacky materials with no adhesion to some plastics. They also cause staining of plasticised PVC.

(b) Polyisobutylene adhesives age better than natural rubber adhesives but other properties are similar.

(c) Vinyl ethers are non-staining but more expensive.

(d) Acrylics apart from an unpleasant smell can have good properties including high temperature resistance.

(e) Silicones are excellent but are very expensive.

All of these materials can be formulated in many different ways so it is not possible to give more than the general outline because good or bad formulation can make nonsense of more detailed information. In testing it is important to look for:

(a) Plasticiser migration from substrate to adhesive or vice versa.

(b) Staining of substrate when subjected to high temperature and/or u.v. light.

(c) Ageing of the adhesive to give non-adhering film. Something to note is that accelerated ageing tests can give misleading results since the high temperature used can cause adhesives to remain tacky whereas at room temperature tack is lost quickly. An example is adhesives based on neoprenes, which are claimed to be permanent tack for polyethylene, which when aged at high temperature remain good but within one month at room temperature have no adhesion to polyethylene at all. This may be due to retardation of crystallisation by the slightly elevated temperature.

(d) Where adhesives are applied from solvent, slight swelling of most plastic films occurs which gives a tunnelling effect.

(e) When used to adhere to metals the adhesives are not strong enough to hold substrates at high temperature as the plastic expands more than the metal.

The main problem with permanent tack adhesives is the low resistance to creep especially when warm.

Where films are to be stuck to either plastic or metal surfaces permanent tack adhesives often give sufficiently strong bonds. A slight amount of adhesion can be given to most surfaces which need not be pretreated and it is sometimes useful to use an adhesive of this type to hold parts together during installation. Even PTFE can be held in this way. On a production line a range of tapes based on permanent tack adhesives supported on non-woven fabrics are very useful for holding light objects before more permanent mechanical fixings are applied. These tapes must be tested for such effects as plasticiser migration because they remain on the job.

Recent advances in this field are: (a) high temperature resistant acrylics, (b) curing polyurethanes, (c) adhesives partially cured on plastic films which have a non-tacky surface layer which when rolled breaks up allowing the lower tacky layers to contact the other substrate.

Choice between one or two way stick

It is normal in production to require the adhesive to take a load immediately the joint is made. Examples are in attaching headlinings, hood sound deadeners and door pad edge folding. It would be ideal to apply the adhesive to one substrate only but because of limitations on tack range two way stick methods should normally be used. When using flexible sheeting as one substrate it is not possible to predict where the adhesive must be and therefore to avoid putting a wide band of adhesive in position a transfer stick is used. Adhesives must be very resistant to creep under load immediately the joint is made.

Where two very different materials are to be adhered

A bond can be produced by using a specially formulated adhesive containing two components each of which has solubility parameter similar to one of the substrates.

The easier method is to prime one surface with an adhesive which adheres to that surface. Another adhesive is then used which will adhere to both the other substrate and the first adhesive. The classic example being the adhesion of nylon to steel where a nitrile phenolic is used to prime the steel and a resorcinol formaldehyde then used to produce the bond.

Surface pretreatments

In many cases it is necessary chemically or physically to change the structure of the plastic surface before adhesive bonding. This applies particularly to crystalline thermoplastics.

Methods used are:

(a) oxidative which can either leave an oxidised layer or may remove certain areas of the plastic as in A.B.S.

(b) Electrical discharge which leaves a more reactive surface.

(c) Ionised inert gas which is claimed to give a stronger plastic surface but I feel that a more reactive surface must also be produced.

(d) Metal ion with fluorinated plastics (Na in liq. NH_3).

With all of these methods the effectiveness of treatment decreases with time and it is very important that the etched or treated surface should be used as soon a possible. One way is to use a priming resin. Ultraviolet light has particularly bad effects on some pretreatments e.g. PTFE.

Summing up this section

If a thermoplastic is to be bonded to itself or a similar material either heat or solvent welding should be used, the actual technique depending on material and shape as well as available apparatus. If unlike materials are to be bonded, adhesives are usually necessary and surface pretreatment, effect of solvents and differences in thermal expansion must be taken into account. With thermosetting plastics—thermosetting adhesives should always be used where structural properties are required. Do not overdesign and always use fastest and cheapest method available.

Plastic materials

It is obviously difficult to obtain agreement from everybody on the best materials and methods for use with the various plastics as most people have different requirements for the system. In this section the recommended systems are for the plastic in its most common form and where possible is the most convenient to use in consumer products. The economics will vary slightly from one industry to another.

The variables which must be considered are: (a) Material form (e.g. film etc.), (b) Design of joint, (c) Apparatus available, (d) Time available, (e) Environmental conditions, (f) Strength requirements over complete temperature range.

Thermoplastics

A.B.S.—does not like plasticisers. Can be ultrasonic, hot plate or spin welded. Can also be hot gas welded but is very strained unless annealed at 135°C.

Solvent welded using 10 per cent A.B.S. in 1, 2 dichloroethane, coated on both surfaces and when tacky put together under a pressure of at least 15 lb/in^2. It is several days before effect of solvent passes. Several other solvents can be used.

Best adhesives are nitriles, nitrile phenolics or polyurethanes. Effect of chromic acid is to etch away the polybutadiene phase as shown in Plate 29 (see plates 30 and 31).

Acrylic—does not like plasticisers. Ultrasonic welding over large peripheries can now be carried out using multiple probes.

Solvent welded using either resin, methylmethacrylate monomer and methyl chloride, or, resin, methylmethacrylate monomer and benzoyl peroxide. The latter is very good for gap filling but being two part is more difficult to use.

With sheet, nitrile/PVAc mixtures or polyurethanes are good adhesives, but with mouldings no solvent stronger than petroleum naphtha can be used. Two part polysulphides give good flexible bonds.

Polyacetal—spin welded with peripheral velocity of 20 ft/sec under a pressure of 200–400 lb/in^2. Hot plate, resistance wire, induction and hot gas welding can be used. Dielectric welding can be used but a frequency of 50 megacycles is necessary.

If surfaces are roughened, oxidised or satinised, neoprenes will give bonds of 500 lb in shear and epoxy-polyamides a bond strength of 1000 lb.

Chlorinated Polyether—dielectric welding is possible. Polyurethane adhesives give best results.

Cellulosics—contain plasticiser. Can be heat welded but this is not used with nitrate because of the fire risk. Acetone can be used to solvent weld most cellulosics, but ethyl acetate is used with ethyl cellulose. Most commonly used adhesives are cellulose nitrates, cyanoacrylates and polyurethanes in acetone.

Fluoroplastics—the themoplastic types other than PTFE may be heat or resistance wire welded. PTFE may be welded by heating to about 350°C under pressure with unsintered PTFE in the joint.

Before using adhesives the surface of the plastic must be etched with Na in liquid ammonia or similar mixture. Non-bonded areas can be protected with PVC tape. Epoxides, polyurethanes and cured neoprenes all give good bonds to etched surfaces which have not been exposed to u.v. light.

Ionomers—can be heat welded and films which have been chromic acid or corona discharge etched can be adhered using nitriles and epoxides. Permanent tack adhesives based on PIB can be used to adhere untreated film.

Urethanes—can be heat and dielectric welded. Can be solvent welded using tetrahydrofuran and resin, and with lower molecular weight resins MEK can be used. Best results are obtained with polyurethane adhesives.

Chloro-xylylene (Paralenes)—used as a coating on any plastic or metal by adsorption of monomer on surfaces. It should be possible to heat weld surfaces moistened with biphenyl solvent.

Nylon—Hot plate, spin and induction welding are all possible but temperature depends on the type. Ultrasonics can also be used if probe is near the joint. No solvents for this class of plastic are satisfactory for welding. Formic acid has a bad long term effect on plastic.

Adhesives are:

(a) 12 per cent aqueous phenol, clamped together and placed in boiling water for 5 min;

(b) 2:1 Calcium chloride to nylon in 6 parts ethanol, filter, coat both surfaces and hold under contact pressure;

(c) 9 per cent phenol in epoxy-polyamide has been used to bond nylon to metal;

(d) Resorcinol-formaldehydes can be used to bond nylon to nitrile phenolic coated metal.

Adhesives become less effective as the nylon number increases.

Polybutylene—not heat sealed because of crystalline changes. Can be solvent welded using chlorinated solvents under heat and pressure. If corona discharge etched, polyurethanes and other rubber based adhesives give good results.

'*Phenoxy*'—is itself a good hot melt adhesive for metals and can be used for metal laminating. Can be hot plate and induction welded to itself and to metals. Can be solvent welded using a 40–50 per cent solution in M.E.K. Isocyanates can be added to bond to other substrates. Adhesives not usually necessary, but solvent based rubbers can be used.

Polyolefins—most forms of heat welding may be used but not dielectric. Ultrasonic is effective only on harder types.

Chromic acid or corona discharge will give adhesive bondable surfaces as will flame treatment. Hot melts are quite effective. Polypropylene primers will give 400 lb/in^2 in shear with epoxides but peel is very poor.

Chlorinated Polyethylene—As for polyolefins with low chlorine content. Higher chlorine contents can be dielectric welded.

Ethylene vinyl acetate—is a hot melt in its own right. Can be heat welded to itself or to other plastics or metals. Can also be dielectric welded.

Polycarbonates—Can be hot gas welded using nitrogen at 650°C and can be hot plate, ultrasonic or spin welded.

A 2:1 mixture of methylene chloride and ethylene chloride thickened with polymer is commonly used, but straight methylene chloride is effective for small areas. At least 200 lb/in^2 pressure is recommended. It gives a white bond if too much solvent or too little pressure is used.

Epoxy polysulphides and solventless polyurethanes can be used. Polyester and solvent based polyurethanes are good but may stress crack mouldings.

Polyethylene Terephthalate—film forms, Mylar or Melinex can be heat welded if benzyl alchohol is used to moisten the joint. Adhesives are effective on corona discharge treated surfaces.

Moulded forms such as Arnite can be spin, hot plate and hot gas welded. Adhesives not yet developed but sodium hydroxide solution will etch the surface.

Polyphenylene Oxides—can be ultrasonically welded in 2 sec with pressure of at least 25 lb/in^2 or spin welded with pressure of 300 lb/in^2. Can also be hot plate welded. Can be solvent welded using chloroform, ethylene dichloride or toluene.

With adhesives stress cracking is a severe problem, but neoprenes and nitriles can be used on low stress parts. After chromic acid pretreatment epoxide adhesives are very effective.

Polysulphones—stress crack easily. Can be spin or hot plate welded and ultrasonically welded if horn is near joint. Can be solvent welded using chlorinated hydrocarbons. Styrene polyesters and resorcinol formaldehyde adhesives can be used.

Polystyrenes—tend to stress crack and are badly affected by plasticisers. Can be ultrasonic, impulse, hot plate or spin welded; can be solvent welded using polymer dissolved in MEK or toluene with solvent naphtha to slow down drying rate. Trichlorethylene can also be used.

Because of attack of solvents on polybutadiene phase and subsequent stress cracking, parts should be annealed before using solvent based adhesives such as neoprenes and nitriles. Styrene polyesters and solventless polyurethanes give good results on stressed parts.

S.A.N.—gives very good results with ultrasonic welding and can also be hot plate and spin welded. Best adhesives are styrene polyesters and polyurethanes.

M.B.S.—Treat as A.B.S.

Polycarbonate/A.B.S. Alloys—Treat as A.B.S.

Polyvinyl Alcohol—can be heat sealed if surfaces are moistened with water and can be solvent welded using water with about 20 per cent glycerol added.

Polypropylene Oxide—similar to polyvinyl alcohol.

Polyvinyl Chloride (*rigid*)—includes various copolymers but mainly with polyvinyl acetate. Both hot gas and dielectric welding is effective.

Can be solvent welded using tetrahydrofuran and resin or cyclohexanone, but, slight heat is required to give good bond.

Adhesives which can be used are:

(a) curing or non-curing nitrile phenolics;
(b) polyurethanes;
(c) neoprenes or S.B.R.s where staining is not a problem;
(d) acrylic or vinyl ether permanent tack adhesives.

Rigid PVC often contains up to 5 per cent plasticiser, this is sufficient to affect bonds produced with epoxide and other non-rubber types.

Polyvinyl chloride (*plasticised*) *and copolymers*

The incorporation of from 10–35 per cent plasticiser gives PVC a rubbery character; because of this it is ideal in sheet form as a covering material and it is normally stretched into position to give a crease free surface. Although stress relaxation does occur there is considerable immediate strain on the bond. The PVC can be supported or unsupported. I have already discussed various problems such as staining of PVC and adhesive failure due to solvent absorption. The only other point to stress is that ageing tests must be carried out to ensure that the adhesive does not absorb plasticiser and soften or even dissolve.

A point to consider when required to adhere PVC to another plastic is the effect of the PVC plasticiser on the other substrate. For example, it will craze and dissolve some polystyrenes, it will stick to ABS if heated, it will craze stressed mouldings of acrylics, polycarbonate, polysulphone and polyphenylene oxide. If something goes wrong the adhesive will be blamed.

Adhering must: (a) give good immediate grab; (b) be non-staining; (c) be plasticiser resistant; (d) not swell the PVC too much.

Because of swelling which is caused by most solvents it is preferable to use heat reactivation, and this is used in both lamination and vacuum covering. Where cover is only adhered round the edge the solvent based adhesives should be applied on the back so as to give shear forces instead of peel.

Adhesives are best based on nitrile or polyurethane. Where long tack range is required neoprenes may have to be used and it is important to obtain good plasticiser and staining resistance.

Because many operations are carried out by hand it is necessary that the adhesive used should be formulated to give exactly the right tack properties. This can save considerable time on a production line.

The other bonding system used with PVC is dielectric welding which is most likely at its best on this material. PVC is bonded to itself with or without automatic cutting. It can be bonded through padding, which should be PVC foam for best results, although treated polyurethane and non-woven fabrics are used.

It can be bonded to hardboard etc. by coating the hardboard with a bonding agent based on PVC latex.

The bonding agent must:

(a) soak into the hardboard surface to slight extent,

(b) coalesce at bonding temperatures,

(c) resolidify immediately power is removed.

One final point on PVC is that a wide variety of plasticisers are used, and if the source of supply is changed it is wise to check adhesive.

Thermoset plastics

By their very nature thermosetting plastics can not be heat or solvent welded and therefore when it is necessary to join a thermoset to itself or to another substrate an adhesive must be used. Because of their greater heat resistance they can more easily be bonded using heat cured systems which give greater bond strength due to the greater cohesive strength of the adhesive.

The surface preparation of thermosets has already been covered. The abrasion treatment also has the effect of bareing the reinforcing medium in reinforced thermosets; it is often easier to bond the reinforcement than to bond to the plastic itself.

Diallyl Phthalate Sanded surface is required, this will give excellent bonds using diallyl phthalate modified polyester after high temperature cure, or epoxy (polyamide or polysulphide) at room temperature. Very good bonds can be obtained.

Epoxides The adhesive should be applied to sanded surfaces and should be of a similar nature to the base resin. Strength obtainable is similar to that of the base resin. In general the curing temperature should be as high as possible allowing for the distortion of the substrate. In this way maximum strength and heat resistance is obtained.

Fast bonding nitrile phenolics with curing resins can give excellent bonds if cured under pressure at temperatures of 300°F. When bonding to metal it is worth noting that many curing nitrile phenolics of the above type will bond to oiled metal surfaces to give bond strengths of about 1000 lb/in^2 in shear in a thin bond line. Bonds of lesser strength can be produced with most rubber based adhesives.

Polyesters Bonded preferably with polyesters of similar composition to substrate, adhesion is good to many bare clean metals. Epoxy/polyamide can also be used. Good bonds can be produced for structural purposes using nitrile phenolics and phenolic epoxys.

Phenolics With many types of phenolic moulding compounds the best bonding medium is resorcinol formaldehyde adhesive. Release agents can be a problem, especially the stearates which are included in some phenolic formulations. Phenolic adhesives give good results but require higher temperatures for cure and are less water resistant than resorcinol based.

Urea, Furan and Melamines All give excellent bonds when using sanded surfaces. Adhesives can be either epoxys or nitrile phenolics. High temperature cured epoxy phenolics and nitrile phenolics are best.

Allyl Diglycol Carbonate Used for optical purposes can be adhered using the monomer with 3 per cent benzoyl peroxide added. The adhesive can be thickened to required viscosity by heating to 50°C in a water bath. It should be refrigerated if not required immediately. Heat to 70°C to cure. The adhesive is also useful with glass lenses.

Silicones Should be adhered with silicones adhesive only. Primers are obtainable for most other surfaces or use moisture curing types.

Polydiphenyl Oxides Thermosetting varnishes are sometimes used to produce laminates with very good heat resistance. To adhere these laminates the varnish which normally contains a heterogenous catalyst can be used. The best method is to apply the solvent based varnish to the surface and allow it to dry. It becomes a rubbery solid which will act as a curing film adhesive when heated to 200°F. The varnish can also be coated on to glass cloth and dried, then cut to required shape for use.

Polyimides Polyimide adhesives not normally obtainable, the varnishes can only be used as adhesives if other substrate is very porous. Otherwise, being a condensation process, condensation products cause the insoluble resin to form as a powder rather than a coherent film. At sufficiently high pressure it should be possible to produce a bond especially if a breather cloth is included.

Nitrile phenolics cured at high temperature will adhere to these materials but the heat resistance is so low compared to the polyimides it would not seem logical to use them. Maximum heat resistance for long term is about 200°C.

Polybenzimidazoles The isophthalamide type can be used as an adhesive for both itself and metals. Cure must be under high pressure (about 200 lb/in^2). A high filler concentration improves breathing of the condensation products. Cure at 600°F.

Polybenzothiazoles, *Polyimide-Amides* and *Polyesteramides* are similar to above materials and have some problems in use.

Foams

Many adhesives can be used because strength of foam is not sufficiently great to require high strength bonds.

Polystyrene Solvent used must not collapse the foam. Water based material is often used. Spray on neoprenes often used industrially.

Polyurethanes—Flexible Butyl or butyl nitrile mixtures can be used. Effect of staining of PVC covers must be considered. *Rigid* Epoxide adhesives often used. Polyurethanes very effective.

Conclusion

The future for plastics is very bright indeed and it is important that the art of plastics bonding should become a science so that a possibly bright future becomes a certainty. Much more work is needed on the theoretical basis of plastics bonding to bring it into line with the available knowledge on rubbers and metals. I realise that much work is being carried out but I would say that the most pressing requirements of the automotive industries are:

(a) improved high speed methods of surface pretreatments to give structural bonds for the more widely used 'cheap' plastics.

(b) development of adhesives with extended open time which cure to give high temperature resistant bonds without heat or pressure. For example, for trim use I consider a permanent tack adhesive with good resistance to cold flow, which then cures to give a strong heat resistant bond, would replace all other types if available.

(c) a physical bonding method for thermoset plastics in order to speed production. For example, polyester based polyurethane foams may be decomposed by heat and the decomposition products re-react to give a bond to another substrate, normally cloth.

With a very high energy electron beam apparatus and the plasma flame it could be possible that even thermosetting plastics may be heat welded. As far as I know, electron beam investigations have never been carried out on plastics; perhaps they should be.

It has been predicted that the growth rate for plastics will be 15 per cent per year up to 1975; this is more than 3 times the expected growth rate in manufacturing generally. 350 per cent increase is expected in plastics usage in the automotive industry to a total of 100 lb per car.

I hope that this paper although not illustrating any breakthrough in bonding technology will encourage others to look more deeply at this field.

Acknowledgement

I would like to thank Ford Motor Company for permission to present this paper.

An Introduction to Surface Energy

PROFESSOR D. J. ALNER, M.Sc., Ph.D., F.R.I.C.
THE CITY UNIVERSITY, LONDON

This lecture is intended to provide an introduction to the concept of surface energy; its purpose is to give in the simplest terms an outline description of the fundamentals of the subject and to relate these to the adhesional process.

Forces between molecules are considered to be universal in character and definitions of these forces have been dealt with at earlier conferences. It seems to be generally accepted that provided sufficiently intimate molecular contact is achieved at an interface, attractive forces of the London dispersion type should by themselves, be sufficient to enable a good adhesional performance to be observed. If we accept the existence of these universal intermolecular forces it follows that strong interfacial attractions must exist at all small distances of intermolecular separation, regardless of the chemical nature of the phases in contact. Thus it has been calculated that if two perfectly plane parallel plates are in molecular contact, i.e. when the centres of the molecules existing in the surface layers are separated by a distance of about 4 Å, the attractive forces should be of the order of 10^9–10^{10} dyne cm^{-1}. Naturally the magnitude of such forces will also depend to some extent on the materials themselves, but it nevertheless follows that poor adhesion must normally be associated with a limited interfacial contact.

When we are dealing with liquid adhesives the problem of achieving intimate molecular contact involves the wetting of the solid surface by the liquid adhesive. It is a well-known fact that a thin layer of liquid which completely wets two flat solids can serve as an adhesive and was demonstrated by Budgett[1] who showed that two highly polished steel plates of 4·5 cm diameter when completely wetted by a film of paraffin oil having a surface tension of 28–30 dyne cm^{-1} required a force of about 20 kg to pull them apart. Even when complete wetting is not apparent it is nevertheless often still possible to get good adhesive action by merely pressing together until only a thin layer remains.

Such methods of forming an adhesive joint have obvious limitations because the shear strength of the liquid is low, being governed by the viscosity of the liquid film. A stronger and more useful joint results if the viscosity of the liquid is greatly increased, e.g. by solvent evaporation, polymerisation, or solidification, thereby providing good tensile and shear resistance. Thus McBain and Hopkins[2] reported that 'any fluid which wets a particular surface and is then converted into a tenacious mass by cooling, evaporation, oxidation, etc., must be regarded as an adhesive for that surface'. Shortly afterwards was added the additional requirement that the adhesive must also be able to deform during the 'setting' process to release elastic stresses developed in the formation of the joint.[3] Their work defined the three essential requirements of an adhesive (1) wetting, (2) setting, (3) deformability to release elastic stresses. In relation to the aspects of surface energy dealt with in this paper, it is the wetting factor which is the important one.

Surface energies of liquids

It is believed that a free surface energy arises at the boundary of a liquid because of an imbalance of intermolecular attractive forces. These arise from the universal London dispersion forces, and in the case of polar molecules from additional dipole-dipole attractive forces. For hydrocarbons, in which only dispersion forces are likely to operate, typical free surface energies are of the order of 15–30 dyne cm^{-1} (e.g. n–hexane 18·4); polar liquids, in which hydrogen bonding between the molecules can occur, have surface energies lying between 30–70 dyne cm^{-1} (e.g. water 72·8). In the somewhat special case of fused salts, which possess aditional large ion-ion attractive forces, surface energies of about 100–600 dyne cm^{-1} are found, and for liquid metals values of 500–5000 dyne cm^{-1} are encountered.

Whenever one distinct phase adjoins another, conditions exist at the boundary of separation which are different from those which occur in the general mass of either of the homogeneous phases bounding it. Consider the surface of a liquid in contact with its vapour: a molecule in the interior of the liquid is subjected to intermolecular attractive forces, but being surrounded on all sides by neighbouring molecules it is attracted equally in all directions (see Fig. 79). However, a molecule in the surface of the liquid is subjected to a net attractive force directed inwardly towards the bulk of the liquid which is not compensated by an equal attraction from the more highly

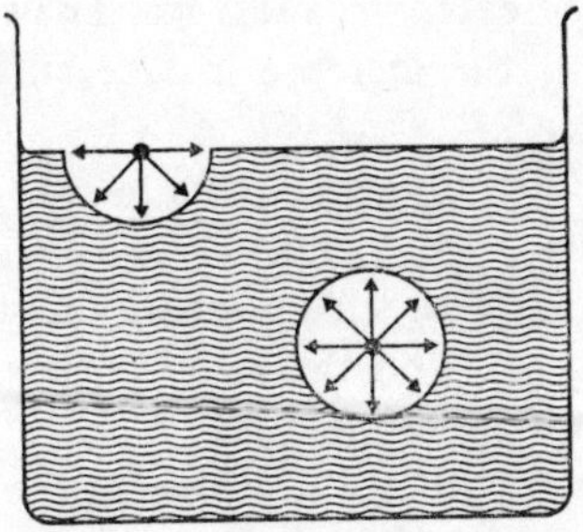

Fig. 79. Molecular forces in a liquid (taken from *Physical Chemistry*, p. 369, by F. T. Gucker and R. L. Seifert) (*Courtesy English Universities Press and Norton and Company Inc.*)

dispersed vapour molecules. Consequently molecules in the surface layer are in a one sided force field and the resultant inwardly directed force tends to draw molecules from the surface into the body of the liquid (see Fig. 80). This results in a greater intermolecular distance of separation between the molecules remaining in the surface layers in directions parallel with the surface. Molecules pulled apart from their equilibrium positions to a greater distance of intermolecular separation increase in potential energy and hence the liquid behaves as if it possessed an elastic skin with a tendency to contract.

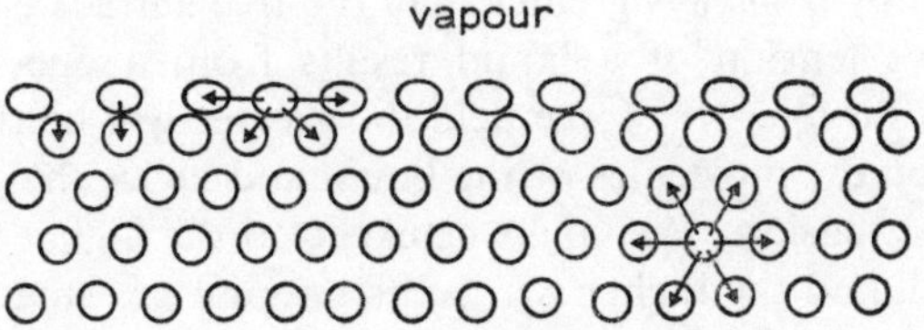

Fig. 80. (Taken from *Treatise on Adhesion and Adhesives*, Vol. I, p. 327, by R. L. Patrick) (*Courtesy Edward Arnold and Marcel Dekker Inc.*)

To extend the area of the surface, i.e. to create fresh surface, molecules must be transferred from the interior of the liquid to the surface, and in so doing work must be done in overcoming the resultant inwardly directed molecular forces; thus these molecules are raised to a state of greater potential energy than those in the bulk of the liquid. This extra energy is usually described in terms of a surface tension acting parallel to the surface of the liquid which opposes any attempt to extend the surface. The surface tension of the liquid, γ, is defined as the increase in free surface energy, $d\gamma$, per unit increase in surface.

$$\gamma_L = \frac{\Delta G_\gamma}{\Delta A} = \frac{\text{dynes} \times \text{cms}}{\text{cm}^2} = \text{dyne cm}^{-1}.$$

Thus for liquids, specific free surface energy and surface tension can be identified as being one and the same property and is a quantity which can be measured experimentally.

The free surface energy does not represent the whole of the energy expended when a fresh surface is formed. The molecules in the surface layer are less densely packed and less well organised than they are in the bulk of the liquid and so it might be argued that the formation of a surface always brings about an increase in entropy and one would expect the free surface energy, γ, to be less than the total surface energy, h, as represented by the expression

$$\gamma = h - Ts$$

from which it follows

$$\gamma = h + \frac{T\,d\gamma}{dT}$$

Total surface energy may be determined by measurements of the free surface energy γ and of the temperature coefficient of this quantity $\frac{d\gamma}{dT}$. Since $\frac{d\gamma}{dT}$ is always negative for pure liquids the surface enthalpy is always greater than the free surface energy.

The surface tension of a liquid results from a one-sided force field and the stronger this force field the greater will be the cohesive forces between the molecules of the liquid and hence the greater will be the surface tension. As might be expected the higher the surface tension of a liquid the higher will be its internal pressure.

Internal pressure is the resultant of the forces of attraction and repulsion operating between the molecules of the liquid and under normal conditions is a net attractive force. This must obviously depend upon the volume V and in turn upon the external pressure P. Thus increase in applied pressure must also produce a marked decrease in the surface tension of the liquid.

Surface energies of solutions

The surface phase of a pure liquid has the same composition as the bulk phase. For solutions, however, the composition of the surface is not necessarily identical with that of the underlying solution. In general the addition of a solute to a solvent will affect the surface tension of the latter and solutes that lower the surface tension of the solvent will tend to accumulate at the surface, whereas solutes which raise the surface tension of the solvent, will tend to accummulate in the bulk phase.

The expression relating the excess or deficiency of solute in the surface phase is given by the Gibbs adsorption equation

$$\Gamma = \frac{1}{RT} \cdot \frac{d\gamma}{d \ln X}$$

where Γ is the surface concentration of solute in a solution of mole fraction of solute X.

For dilute solutions, this becomes

$$\Gamma = \frac{1}{RT} \cdot \frac{d\gamma}{d \ln C}$$

Interfacial tensions between liquids

(a) *Non-polar liquids* When two immiscible liquids are in contact they are separated by a definite boundary or interface and the two adjacent interfacial monomolecular layers are both subjected to an unsymmetric force field. Consequently both have a tension the sum of which is the interfacial tension. In the case of liquids this can be experimentally measured (e.g. by the drop weight method).

Let us consider the boundary between, say, mercury (A) and a liquid hydrocarbon (B). In the interfacial region on the hydrocarbon side of the boundary, the surface molecules of the hydrocarbon will be attracted towards the bulk of the hydrocarbon phase by the intermolecular forces which cause the surface tension of the hydrocarbon γ_B. The hydrocarbon molecules at the interface will also be attracted by the dispersion attractive forces of the mercury molecules at the interface. Thus the hydrocarbon molecules at the interface will be in a force field different from that existing at the surface of the hydrocarbon not in contact with the mercury. The surface tension of the hydrocarbon molecules in this interfacial region will be some function of the surface tension and of the dispersion attractive forces between the molecules of the hydrocarbon and the mercury. There is evidence for believing that, within certain limitations, the dispersion force attractions at such an interface can be accounted for in terms of the geometric mean of the surface tensions of the two liquids in contact i.e. $(\gamma_A^d \gamma_B^d)^{\frac{1}{2}}$, where γ_B^d and γ_A^d refer respectively to the specific free surface energies of the hydrocarbon and mercury arising from purely dispersion force interactions.[4]

Thus the surface tension of the hydrocarbon in the interfacial region will be $\gamma_B - (\gamma_A^d \gamma_B^d)^{\frac{1}{2}}$; and by a similar argument the surface tension of the mercury in the interfacial region will be $\gamma_A - (\gamma_A^d \gamma_B^d)^{\frac{1}{2}}$.

The total interfacial tension, γ_{AB}, which is the sum of the tensions in the two layers will then be represented by

$$\gamma_{AB} = \gamma_A + \gamma_B - 2(\gamma_A^d \gamma_B^d)^{\frac{1}{2}}.$$

The use of this geometric mean relationship makes certain assumptions about the sizes and shapes of the molecules and also assumes that only dispersion force attractions operate in the interfacial region. Certain cases are known to approximate to these requirements and for these the relationship appears to explain the observed data.

The expression has been tested with a number of liquid systems, one of which has been hydrocarbons and water. In the case of hydrocarbons it can reasonably be assumed that $\gamma_B \equiv \gamma_B^d$ since dispersion forces only are known to operate in this phase and hence the only unknown is that of $\gamma_{H_2O}^d$ and a set of typical data for the water-hydrocarbons system is shown Table 24.

TABLE 24

Water-Hydrocarbons at 20°C. $\gamma_{H_2O} = 72{\cdot}8$ dyne cm^{-1}. *

Hydrocarbon (B)	γ_B	γ_{AB} (measured)	$\gamma_{H_2O}^d$ (calculated)
n-hexane	18·4	51·1	21·8
n-heptane	20·4	50·2	22·6
n-octane	21·8	50·8	22·0
n-decane	23·9	51·2	21·6
n-tetradecane	25·6	52·2	20·8
Cyclohexane	25·5	50·2	22·7
Decalin	29·9	51·4	22·0

*Taken from *Treatise on Adhesion and Adhesives*, Vol. I, p. 347, by R. L. Patrick. *Courtesy Edward Arnold and Marcel Dekker Inc.*

The constancy of the calculated values of $\gamma_{H_2O}^d$ seems to provide good evidence that the equation satisfactorily explains the observed behaviour of these systems and that the assumption that only dispersion force attractions operate at the interface is essentially correct. Similar observations have also been made with other systems and by this means data has been accumulated about the γ^d contributions of various liquids.

Such data can be used to predict the magnitude of interfacial tensions between any known pair of immiscible liquids, e.g. using the values of $\gamma^d_{H2O} = 22{\cdot}0$ dyne cm^{-1} and $\gamma^d_{Hg} = 484$ dyne cm^{-1} and assuming that only dispersion forces operate, it can be calculated that the interfacial tension arising between these two liquids would be 424 dyne cm^{-1}. This calculated value agrees well with the measured value of the interfacial tension between mercury and water, viz: 426 dyne cm^{-1} and the result appears to indicate that the interactions between water and mercury are in fact almost entirely due to dispersion forces, the contribution from other forces being very small indeed.

(b) *Polar liquids* By means of the relationship $\gamma_{AB} = \gamma_A + \gamma_B - 2(\gamma^d_A\,\gamma^d_B)^{\frac{1}{2}}$ it has been shown how it is possible to evaluate the γ^d terms for any liquid on the assumption that only dispersion forces operate, and thus to compare these calculated values with the energies observed from measurements of interfacial tension. It is thus possible in terms of the deviations between the calculated and measured data to arrive at some assessment of the magnitudes of other interactions involved (e.g. polar forces) when polar liquids are used. Table 25 shows data obtained for the interface between organic liquids and water at 20°C.

TABLE 25

Energy of water-organic liquid interactions in excess of dispersion force interaction at 20°C.

($\gamma^d_{H_2O}$ = 22 dyne cm^{-1}; γ_{H_2O} = 72·8 dyne cm^{-1}). *

Liquid (B)	γ_B	γ_{AB} (measured)	γ^d_{AB} (calcd.)	$(\gamma^d_{AB}-\gamma_{AB})$	$2(\gamma^d_A\,\gamma_B)^{\frac{1}{2}}$
Benzene	28·9	35·0	51·5	+16·5	50·2
Toluene	28·5	36·1	51·5	+15·4	50·0
Mesitylene	28·8	38·7	51·4	+12·7	50·1
Carbon tetrachloride	26·9	45.0	51·3	+ 6·3	48·4
Carbon tetrabromide	49·7	38·8	56·5	+17·7	66·0
Butyronitrile	28·1	10·4	51·3	+40·9	49·6
Aniline	42·9	5·8	53·7	+47·9	62·0

*Taken from *Treatise on Adhesion and Adhesives*, Vol. I, p. 349, by R. L. Patrick. Courtesy *Edward Arnold and Marcell Dekker Inc.*

In this table values for γ_B^d (liquid hydrocarbon) have been assumed to be equal to γ_B. For each polar liquid the measured γ_{AB} is shown in the second column. Column 3 shows the interfacial tensions γ_{AB}^d calculated on the assumption that only dispersion forces are operating. The difference between these two quantities (calculated and measured) are shown in the penultimate column.

If polar attractions in excess of dispersion force attractions operate the interfacial tension will be reduced as represented by

$$\gamma_{AB} = \gamma_A + \gamma_B - 2(\gamma_A^d \gamma_B^d)^{\frac{1}{2}} - \text{(polar attractions)}$$

and hence γ_{AB}^d (calc.) $- \gamma_{AB}$ (measured) will for such cases, give a positive deviation, as is observed. If no polar forces operate the difference will be zero. It is significant that the dispersion force interaction term represents by far the largest contribution to interfacial tension, except in those cases where hydrogen bonding can occur, e.g. aniline, butyronitrile.

Solid-liquid interfaces

For liquid-liquid systems interfacial tensions, which are capable of direct measurement, can be identified with free surface energies. For solid-liquid interfaces, however, it is not possible to directly measure the free surface energies of the solids and it is also difficult to conceive of a solid surface possessing a surface tension analogous to that of a liquid. Thus the solid-liquid interface is less well defined and less well understood than the corresponding liquid-liquid systems.

Attempts to explain solid-liquid systems have focused attention on the problem of wetting. It is likely that most liquids wet most solids to some extent so that there is always likely to be some adhesion at any liquid-solid interface. In contacting surfaces, wetting

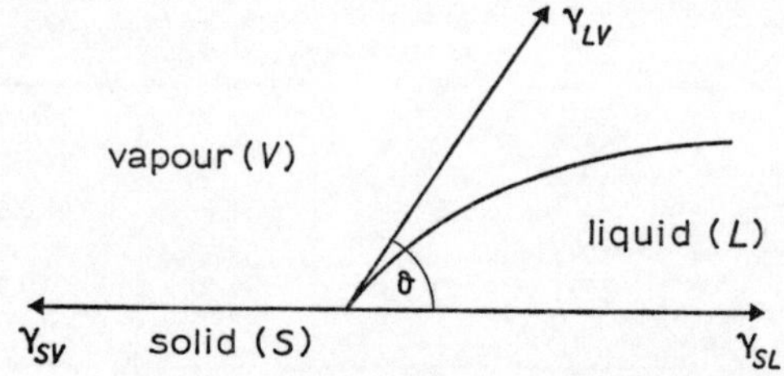

Fig. 81. Vector diagram for liquid drop on solid surface (*Courtesy Royal Institute of Chemistry*)

is of prime importance, and basic to the subject of wettability is Thomas Young's concept of the contact angle formed by a drop of liquid resting on a plane solid surface.[5] If a drop of liquid does not spread completely a state of equilibrium is assumed to arise involving a three phase point of contact between the solid, liquid and vapour phases, as represented in Fig. 81.

Adopting the mathematical convention of treating the free surface energies as vector quantities, the well known Young equation follows, viz.

$$\gamma_{SV} = \gamma_{SL} + \gamma_{LV} \cos \theta \quad . \quad . \quad . \quad . \quad . \quad (1)$$

where subscripts SV, LV refer to the solid and liquid phases in equilibrium with the vapour of the liquid, and γ_{SL} to the solid-liquid interface.

When $\theta > 0$ the liquid is said to be nonspreading but when $\theta = 0$ the liquid is said to wet the solid completely and to spread freely over the surface at a rate depending on the viscosity of the liquid and roughness of the solid surface.

Dupré[6] showed that the reversible work of adhesion, W_A, defined as the (reversible) work required to separate 1 cm^2 of solid-liquid interface, is related to the specific free surface energies by the relation

$$W_A = \gamma_S + \gamma_{LV} - \gamma_{SL} \quad . \quad . \quad . \quad . \quad . \quad (2)$$

where γ_S refers to the bare solid in vacuo. Thus the liquid will attach itself to the solid surface if $W_A > 0$.

Conditions for a liquid to spread on a surface have been defined in terms of a 'spreading coefficient' S, where

$$S = W_A - W_c.$$

W_c is the work of cohesion of the liquid. $W_c = 2\gamma_{LV}$ from which it follows:

$$S = \gamma_S - \gamma_{SL} - \gamma_{LV} \quad . \quad . \quad . \quad . \quad . \quad (3)$$

and spreading will take place when S is positive, i.e. $S > 0$.

A liquid in proximity to a solid can, however, produce an adsorbed vapour film on the surface and the work required to separate 1 cm^2 of liquid from a surface to leave such an adsorbed surface film will be

$$W_A = \gamma_{SV} + \gamma_{LV} - \gamma_{SL} \quad . \quad . \quad . \quad . \quad . \quad (4)$$

where γ_{SV} refers to the free surface energy of the solid with the adsorbed film. Adsorption of such a vapour film involves a decrease in the free surface energy of the solid, i.e. γ_S falls to γ_{SV}. This decrease in free surface energy $(\gamma_S - \gamma_{SV})$ has been termed the spreading pressure, π, and if one assumes the vapour behaves as an ideal gas can be evaluated by means of the Gibbs adsorption equation

$$\pi = \gamma_S - \gamma_{SV} = RT\int_0^p \Gamma \frac{dp}{p}$$

where Γ represents the surface concentration or the number of moles of vapour adsorbed by 1 cm^2 of the solid surface.

The adsorption of a vapour film produces a surface of lower free surface energy. The extent of the decrease in free surface energy may not be sufficient to prevent the liquid attaching itself to the surface, i.e. $\gamma_{SV} + \gamma_{LV} - \gamma_{SL}$ remains positive and > 0, but it may, nevertheless, cause the spreading coefficient, S, to become negative, i.e. $\gamma_{SV} - \gamma_{LV} - \gamma_{SL} < 0$, with the result that the liquid no longer spreads but instead gives rise to a condition involving a finite contact angle θ, as defined by the Young equation

$$\gamma_{SV} = \gamma_{SL} + \gamma_{LV} \cos \theta$$

and since $\gamma_S - \gamma_{SV} = \pi$

then
$$\gamma_S = \pi + \gamma_{SL} + \gamma_{LV} \cos \theta \quad . \quad . \quad . \quad . \quad (5)$$

The free surface energies of most liquids (except liquid metals) are < 100 dyne cm^{-1} at ordinary temperatures. Hard solids have free surface energies ranging from 500 to 5 000 dyne cm^{-1}, the value increasing with hardness and melting point (e.g. metals, metal oxides). On the other hand, most organic solids (e.g. waxes, polymers) have lower melting points and their specific free surface energies are usually less than 100 dyne cm^{-1}. Solids can thus conveniently be classified as high energy or low energy materials, a high energy surface being one which attracts the molecules of the liquid more strongly than they attract one another whereas a low energy surface attracts the molecules of liquid less strongly than they attract one another.

Because most liquids (organic and inorganic) have comparatively low free surface energies, usually, but not always, they spread on high energy solids. However, since the free surface energies of liquids are of the same order of magnitude as those of many low energy solids, non-spreading is most commonly encountered among such systems.

By analogy with liquid interfaces and assuming only dispersion forces operate, the γ_{SL} term can be expressed as

$$\gamma_{SL} = \gamma_S + \gamma_{LV} - 2(\gamma_S^d \gamma_{LV}^d)^{\frac{1}{2}} \quad . \quad . \quad . \quad . \quad . \quad (6)$$

and by substituting for γ_{SL} in Equation (5) we obtain

$$\gamma_{LV} \cos \theta = 2(\gamma_S^d \gamma_{LV}^d)^{\frac{1}{2}} - \gamma_{LV} - \pi \quad . \quad . \quad . \quad . \quad . \quad (7)$$

This expression has been used to assess γ_S^d values from contact angle measurements for a series of reference (hydrocarbon) liquids on low energy surfaces and assuming that $\pi = 0$; this latter assumption being based on the premise that adsorption of a high energy liquid by a low energy solid does not take place.

Rearranging Equation (7) and assuming $\pi = 0$, gives

$$\cos\theta = 2(\gamma_S^d)^{\frac{1}{2}}.\frac{(\gamma_{LV}^d)^{\frac{1}{2}}}{\gamma_{LV}} - 1$$

and a plot of $\cos\theta$ versus $(\gamma_{LV}^d)^{\frac{1}{2}}/\gamma_{LV}$ should produce a straight line with an intercept at -1 and slope $2(\gamma_S^d)^{\frac{1}{2}}$. A reasonable degree of success has been achieved for a number of such systems and typical values for γ_S^d obtained by this method are: paraffin wax 25·5, polyethylene 35 and polystyrene 44.

γ_S^d values have also been derived for a number of high energy surfaces from spreading pressure measurements. Under conditions where multilayer adsorption occurs and is complete, i.e. when the vapour pressure in equilibrium with the adsorbed film is equal to the saturation-vapour pressure, p^0, of the adsorbed liquid, the outer layer of such a multilayer film could be expected to have the same free surface energy as the liquid itself. In these cases spreading conditions exist (i.e. $\theta = 0$), hence

$$\pi_e = \gamma_S - (\gamma_{LV} + \gamma_{SL}) \text{ and } \pi_e = RT\int_0^{po} \Gamma\,.\frac{dp}{p} \quad . \quad . \quad . \quad (7)$$

Substituting $\gamma_{SL} = \gamma_S + \gamma_{LV} - 2(\gamma_S^d\,\gamma_{LV}^d)^{\frac{1}{2}}$ gives

$$\pi_e = 2(\gamma_S^d\gamma_{LV}^d)^{\frac{1}{2}} - 2\gamma_{LV}$$

which on rearrangement becomes

$$\gamma_S^d = \frac{(\pi_e + 2\gamma_{LV}^d)^2}{4\gamma_{LV}^d}$$

γ_S^d values have also been obtained from measurements of heats of adsorption[8] and heats of immersion[9], and for a number of cases a fair degree of correlation has been observed between the results provided by the different methods.

Finally Zisman[10] and his co-workers have observed, for a series of reference liquids on low energy surfaces, that a plot of $\cos\theta$ against γ_{LV} gives a straight line. The intersection of the extrapolation of this line with the line $\cos\theta = 1$, has been termed the critical surface tension for wetting the solid and is also believed to be equal to γ_S^d.

By contact angle and spreading pressure measurements with polar liquids and using γ_S^d data provided from measurements with reference liquids, it has also been possible to make assessments of the excess forces arising from the interactions of the polar molecules with solid surfaces, in a manner analogous to that used for the liquid systems.[11]

This paper has tried to present briefly and in simple terms, a review of some of the progress which has been made towards a better understanding of the adhesional process through the concepts of surface energy. It has been shown for liquid systems how it has been possible to expand interfacial energy into a series of terms for each type of intermolecular interaction and how this has enabled the contributions of each to be assessed. For liquid systems interfacial tensions can be directly measured and hence can provide some verification of the calculated tensions obtained from the expressions set up. It has been indicated how the concepts developed for liquid systems, have been carried over to solid-liquid systems in an effort to develop an explanation of this more difficult and complex system. It is unfortunate that solid-liquid interfacial energies are not capable of direct measurement, but nevertheless it is apparent that significant progress has been made towards some explanation of such systems, although the position is still far from being resolved and controversy still rages about some of the concepts employed.

References

1. Budgett, H. M., *Proc. Roy. Soc.*, **A86**, 25, 1911.
2. McBain, J. W. and Hopkins, D. G., *J. Phys. Chem.*, **29**, 88, 1925.
3. McBain, J. W. and Lee, W. B., *Ind. Eng. Chem.*, **19**, 1005, 1927.
4. Girifalco, L. A. and Good, R. J., *J. Phys. Chem.*, **61**, 904, 1957.
5. Young, T., *Phil. Trans. Roy. Soc. (London)*, **95**, 65, 1805.
6. Dupré, A., *Theorie Mechanique de la Chaleur*, p. 369, Gauthier-Villars, Paris, 1869.
7. Graham, D., *J. Phys. Chem.*, **68**, 2788, 1964.
8. Isirikyan, A. A. and Kiselev, A. V., *J. Phys. Chem.*, **66**, 210, 1962.
9. Healey, F. H., Chessick, J. J., Zettlemoyer, A. C. and Young, A. C., *J. Phys. Chem.*, **58**, 887, 1954.
10. Zisman, W. A., *Ind. Eng. Chem.*, **55**, 19, 1963.
11. *Treatise on Adhesion and Adhesives*, Ed. Patrick, R. L., Vol. 1, Arnold, London.

The Significance of Surface Energy in Adhesion

C. A. DAHLQUIST, B.Ch.E.

MINNESOTA MINING AND MANUFACTURING COMPANY, U.S.A.

Introductory

Some years ago Bikerman[1] proposed that the measured mechanical strengths of adhesive bonds have no relation to the surface energies of adhesive and adherend, but depend only on the condition of the bond and the mechanical properties of the components of the adhesive. Bikerman pointed out that the principal reason adhesive bonds do not achieve ideal strengths is the existence of weak boundary layers.

Schonhorn and Sharpe[2] reached essentially the same conclusions, but since they investigated bonding to 'low energy surfaces' they recognised the significance of surface energy to the extent that it influenced 'wetting' of the adherend by the adhesive. When the surface energy of the adhesive in its fluid state is greater than that of the adherend in its solid state, wetting is incomplete and an imperfect bond is formed. Presumably we then have a weak boundary layer of air. More recently Schonhorn and Hansen have reported[3] that strengthening of the surface of low energy polymers can be accomplished without increasing the surface energy, and strong bonds are then achieved with adhesives which apparently have higher surface energies than the adherends.*

In the face of the above arguments and recognising that the mechanical work of adhesion, typically 10^4 to 10^6 erg per cm^2, completely swamps the thermodynamic work of adhesion, usually about 50 to 200 erg per cm^2, it would seem foolhardy to attempt to draw any correlations between the mechanical bond strength and the work of adhesion.

Nevertheless, it is well recognised that cohesive energy, which is

*If Schonhorn and Hansen did indeed bond an adhesive to a solid of lower surface energy than itself, an inconsistency arises between this accomplishment and the criterion for adhesion set earlier by Schonhorn and Sharpe.

twice the surface energy, can, under certain conditions, be directly related to mechanical work of rupture, as in the measurement of the surface energy of crystalline solids by cleavage.[4] Calculations of forces of attraction in solids[5, 6, 7] show that significant differences exist in the 'ideal' cohesive strengths of materials, and in the attractive forces across interfaces. These considerations warrant a re-examination of the relation between thermodynamic work of adhesion and mechanical bond strength.

'Ideal' bond strengths

Gardon equates the maximum or ideal strength of an adhesive bond, σ_{max}, and the work of adhesion, W_{as}, as follows:

$$\sigma_{max} = 1{\cdot}64 \times 10^8 W_{as}/(V_a^{\frac{1}{3}} + V_s^{\frac{1}{3}}) \quad . \quad . \quad . \quad (1)$$

Where V_a and V_s are the molar volumes of the adhesive, *a*, and the adherend, *s*, respectively. When *a* and *s* are identical, σ_{max} becomes the cohesive strength and W_{as} can be replaced by 2γ, where γ is the surface energy. Gardon prefers to replace γ by the cohesive energy density using the relationship

$$\gamma = \delta^2 V^{\frac{1}{3}}/16 \quad . \quad . \quad . \quad . \quad . \quad . \quad . \quad . \quad (2)$$

where σ is the solubility parameter (δ^2 is the cohesive energy density). He finds that the measured tensile strengths of 'technical grade unoriented polymers' are quite close to the calculated tensile strengths, ranging from about 40 per cent to about 70 per cent of σ_{max}. This lends validity to Equation (1), though the assumption that the molar volumes of solution thermodynamics and adhesion thermodynamics are equivalent is a moot point.

It must be recognised that these equations take no account of the work of deformation preceding rupture, and presumably should give better agreement between theory and experiment in the glassy state than in the rubbery state.

Huntsberger[8] presents a criterion for interfacial failure derived from interfacial energy considerations and contact angle equilibria. Expressed in the terminology used in this paper, the criterion is that interfacial separation will occur when

$$\phi(\gamma_s/\gamma_a)^{\frac{1}{2}} < 1 \quad . \quad . \quad . \quad . \quad . \quad . \quad . \quad . \quad (3)$$

where ϕ is the Girifalco and Good interaction parameter. He then presents experimental evidence to show that interfacial separation occurs when a polyisobutylene adhesive is peeled from an alkyd resin.

Two significant premises come out of the work of Gardon and of Huntsberger:

(1) Real cohesive and adhesive bond strengths are not necessarily many orders of magnitude lower than ideal strengths, and can, in fact, approach ideal strengths very closely.

(2) Adhesive bond failure is possible from thermodynamic considerations.

Pressure sensitive adhesive bonds

In a previous publication[9] Dahlquist compared the works of adhesion and cleavage stresses in bonds between a pressure sensitive tape and various surfaces. Although a 1 : 1 correlation seemed to be justified by the data, there was a great deal of scatter and no firm conclusions could be drawn. It now seems worthwhile to re-examine peel adhesion data under the premise that true adhesive separation can occur from both mechanical and thermodynamic considerations.

In many respects the pressure sensitive adhesive bond appears to be a poor choice for a study of fundamentals of adhesion. Pressure sensitive adhesives are soft, viscoelastic materials which undergo high deformation when a bond is ruptured, so that the ratio of the mechanical work of adhesion to the thermodynamic work of adhesion becomes very high. By their nature they are usually thought to have low cohesive strength, and to be compounded with low molecular weight tackifying resins and oils which can migrate to the interface and form weak boundary layers. However, this is not always the case, because pressure sensitive adhesives can be high molecular weight polymers, even partially cross-linked, so that no obvious source of weak boundary layer material need be present.

Furthermore, in defence of using pressure sensitive adhesives to study adhesion, the mechanism of separation in a viscoelastic adhesive bond undergoing peel has been elucidated in great detail by the theoretical and experimental work of Kaelble,[10, 11, 12] and the rheology of soft polymers has received much attention, giving us insight into the stress-strain behaviour in the pressure sensitive adhesive bond.

Quality of the pressure sensitive adhesive bond

It has been said of pressure sensitive adhesives that they are slow to wet surfaces, even those of apparent smoothness such as commercial films, because of their high viscosity. Yet, most of the pressure sensitive adhesives investigated by the writer achieved 90 per cent or more of their 'ultimate' adhesion to cellophane or polyester film after a contact time of two minutes when roll down was effected with high local pressure (1 500 gram load on a 1·27 cm diameter steel roller) under such conditions that no air was visibly trapped. 'Ultimate' adhesion was the peel adhesion value measured after 2 weeks of contact under a pressure of 16 grams per cm^2. Under these conditions the additional 10 per cent increase in peel adhesion took place during the first 48 hours and no significant increase occurred after 1 week.

The rate at which a pressure sensitive adhesive penetrates the pits and crevices in an adherend surface can be estimated by de Bruyne's[13] analysis of the rate of penetration of a viscous adhesive into a cylindrical pit. For the condition where the rate is very slow, such that steady state viscosity prevails and back pressure of the en-trapped air can be ignored, the equation of motion becomes

$$x\frac{dx}{dt} = \frac{r_o^2 p}{8\eta}$$

where X is distance of penetration, r_o is the radius of the pit, P the pressure, and η the viscosity.

Integrating and solving for the penetration in terms of the pit diameter we have

$$\frac{x}{2r_o} = \left(\frac{pt}{16\eta}\right)^{\frac{1}{2}}$$

If t is taken as one week ($6{\cdot}05 \times 10^5$ sec), p as 16 grams (15 700 dyne) per cm^2, and η as 10^7 poises, a typical steady state viscosity for many pressure sensitive adhesives, we find

$$\frac{x}{2r_o} = 7{\cdot}7$$

indicating that very substantial penetration will take place.

The penetration can also be accelerated by raising the temperature. The decrease in viscosity which occurs when the temperature is raised to 65°C from 23°C can be estimated from the Arrhenius equation

$$\eta = A \exp\left(\frac{E}{RT}\right)$$

where E is the activation energy for viscous flow, R is the gas constant, and T the absolute temperature. The viscosity change, expressed as a ratio, becomes

$$\eta_{296}/\eta_{338} = \exp\left(\frac{E}{R}\left(\frac{1}{296} - \frac{1}{338}\right)\right)$$

A typical value of E for soft elastomeric polymers is 10 kcal per mol in this temperature range. η_{296}/η_{338} then will be 8·3, showing that approximately the same ratio of time to viscosity is maintained when the test condition is one day at 65°C as when it is one week at 23°C. In either case the penetration will be substantial, and this conclusion is supported by the experimental observation that little additional adhesion build-up occurs after one week at room temperature or after 16 hours at 65°C.

Penetration of a fluid adhesive into surface depressions necessitates favourable wetting conditions as well as favourable rheology. The relationships involving surface topology, surface energies, and the degree of penetration of the adhesive into the surface have also been discussed by de Bruyne.[13] Such considerations, of course, require knowledge of the surface energies of the adhesives and the adherend. The writer is not aware of any published work on direct measurement of the surface energies of pressure sensitive adhesives, nor, for that matter, of published data on the critical surface tension as measured by the Zisman[14] technique. Data obtained in our own laboratories indicate that the critical surface energy, or, more properly, the dispersion component of the surface energy (γ^D) as defined by Fowkes[15] and determined by the method of Kaelble and Uy[16, 17] is quite low, varying from about 17 to about 30 erg per cm^2. By equating the work of adhesion from the contact angle equilibrium

$$W_a = \gamma_a(1 + \cos\theta)$$

and the root mean square approximation for the work of adhesion for interfacial separation

$$W_a = 2(\gamma_a^D\, \gamma_s^D)^{\frac{1}{2}}$$

(ignoring for the moment the polar force contributions which are small in typical pressure sensitive adhesives and in low energy surfaces) one calculates that the contact angle of a pressure sensitive adhesive ($\gamma_a^D \cong 30$ erg per cm^2) on polytetrafluoroethylene ($\gamma_s^D \cong 18$ erg per cm^2), is 57°. De Bruyne[13] has shown that some penetration of a pit will occur if the sum of the angle of the pit wall with the horizontal plane and the contact angle of the adhesive on the adherend does not exceed 180°. He also defines a smooth surface, such as the surface of a smooth plastic film, as one in which the irregularities consist only of shallow conical or spherical impressions. Thus it is very possible, and, indeed, likely, that a pressure sensitive adhesive will penetrate and wet a smooth surface, even that of polytetrafluoroethylene, when subjected to pressure and held for sufficient time for air to diffuse away.

It would seem that penetration of cylindrical pits and displacement of air in shallow depressions should be sufficient criteria for wetting, but if 'wetting' is arbitrarily defined as the condition where γ_s exceeds γ_a, or γ_s^D exceeds γ_a^D, then the above statement is invalid. However, we also have situations where liquids display a low or a high contact angle on solids depending upon the relative values of γ_L and γ_s, and it seems that we should then talk about 'degrees of wetting'. A more important consideration is whether or not the population density of defects in a pressure sensitive bond increases as the surface energy of the adherend decreases. It would seem that this will not be the case if the pressure sensitive adhesive bond is held under pressure for a sufficient and not unreasonable length of time.

Mechanism of bond rupture

Kaelble[12] has carried out both theoretical and experimental analyses of the stress distribution in a pressure sensitive adhesive bond undergoing peel. The model which he chose for the theoretical analysis assumed that the maximum tensile stress was reached at the bond boundary. The actual tensile stress distribution was measured by a 'split beam' instrument which sensed the compressive and tensional load as the peeling zone approached and passed over the end of a cantilever beam in close proximity to a matching beam which formed a coplanar adherend surface for the bond. The stress distribution was obtained from the slope of the force-displacement curve (Fig. 82).

The stress distribution measured in this fashion is an apparent stress distribution. The bond stress analyser senses the total stress across the width of the bond, but this is not true stress (i.e. stress per unit of cross-sectional area of stretched adhesive) because the adhesive breaks up into strands. The experimental stress profile shows that the maximum apparent tensile stress is attained some distance ahead of the bond boundary. However, the strands become highly extended

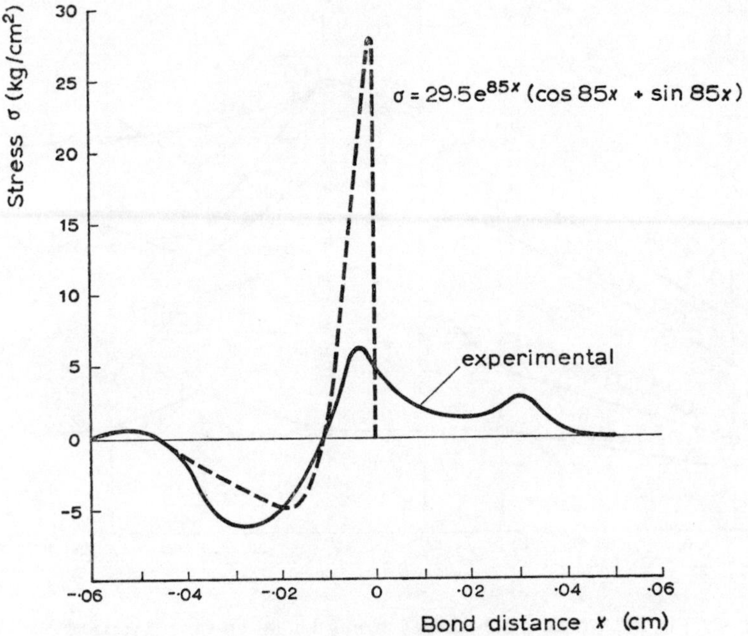

Fig. 82. Peel adhesion[18]

before separating from the adherend (see photograph in Ref. 18). The extension of a strand causes a corresponding decrease in its cross-sectional area, and the true stress becomes correspondingly greater than the apparent stress. In the specific example of Fig. 82, the apparent boundary stress is about 2·4 kg per cm², but the adhesive strands are extended about 10 fold, and the true stress is estimated to be 10 times greater, or 24 kg per cm². This is in good agreement with the 25 kg per cm² boundary stress which Kaelble calculates from his equation of peel.

Under the conditions of the above experiment we find that the time of extension of the adhesive is about 5 sec (calculated from the rate of peel, 0·5 cm per minute, and the distance from the point of zero stress to the point of maximum true stress, about 0·04 cm). Stress-strain curves were run on several adhesives, including the kind of natural rubber-resin adhesive used by Kaelble. Under the test conditions the rate of loading rather than the extension was controlled, but in general an extension of 1 000 per cent was reached

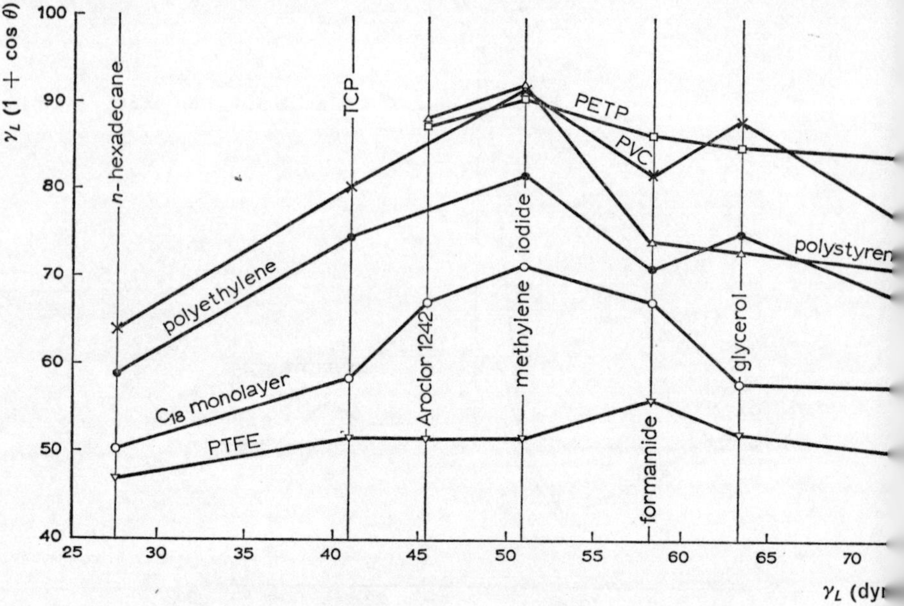

Fig. 83. Works of adhesion of various liquids on various surfaces

in 4 to 6 seconds. Stresses of 25 to 27 kg/cm² were obtained at 1 000 per cent extension for the rubber-resin adhesive. The breaking strength, however, reached 48 to 51 kg/cm² at 1 610 to 1 630 per cent extension. Therefore the maximum stress in the bond is in good agreement with the measured stress-strain behaviour of the adhesive, but is substantially less than the measured ultimate strength. Other adhesives which had higher ultimate tensile strengths did not show proportionately higher peel force values. In fact, they were frequently lower.

Why is it that stress in the adhesive strand does not reach the ultimate strength of the adhesive as it should if the pressure sensitive adhesive fails cohesively? Before we address ourselves to this question let us examine the boundary stress in relation to the work of adhesion.

Work of adhesion in the pressure sensitive adhesive bond

In estimating the work of adhesion between the pressure sensitive adhesive and the adherend, use was made of the relationship proposed by Kaelble and Uy:

$$W_a = 2\,[(\gamma^{L}_{D}\,\gamma^{D}_{S})^{\frac{1}{2}} + (\gamma^{P}_{L}\,\gamma^{P}_{S})^{\frac{1}{2}}] \quad . \quad . \quad . \quad . \quad (4)$$

where γ refers to free surface energies, the subscripts L and S to liquid and solid, respectively, and the superscripts D and P to the dispersion and polar contributions to the free surface energy.

γ^{D}_{S} was taken as γ_c of Zisman,[14] while γ^{P}_{S} was estimated from liquid contact angle measurements. The liquids and the 'dispersion' and 'polar' components of their surface energies are given in Table 26. The works of adhesion of the liquids on the following surfaces, polyethylene, polyethylene terephthalate, polyvinylchloride, polystyrene, polyethylene, oriented octadecyl amine monolayer, polytetrafluoroethylene, and polychlorotrifluoroethylene, were taken from published data[19] (see also Fig. 83). γ^{P}_{S} was chosen to give the best fit between the quantity (Table 28)

$$W_a - 2(\gamma^{P}_{L}\,\gamma^{P}_{S})^{\frac{1}{2}}$$

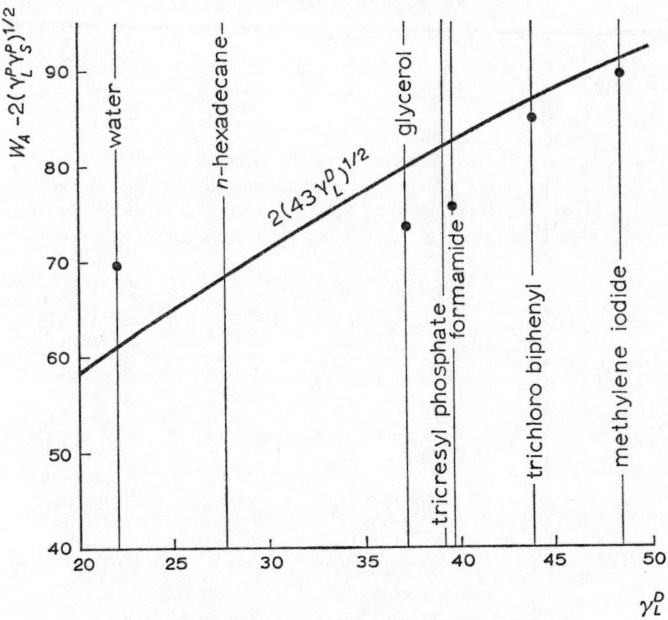

Fig. 84. Wetting of polyethyleneterephthalate

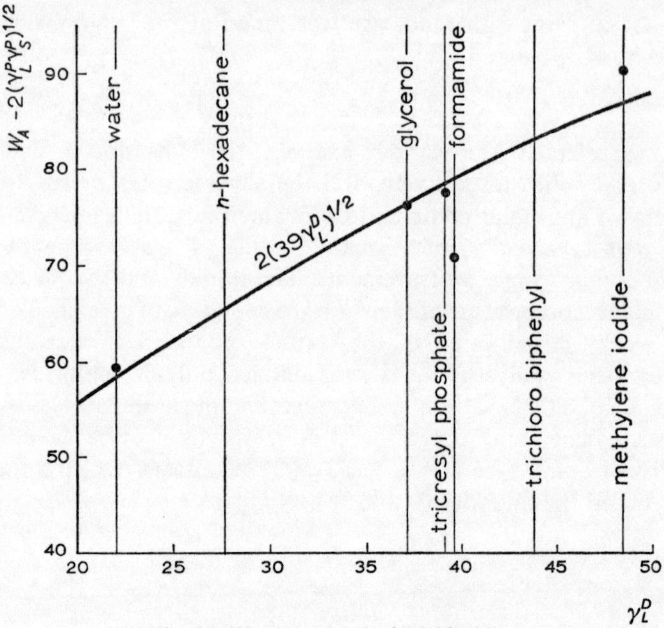

Fig. 85. Wetting of polyvinylchloride

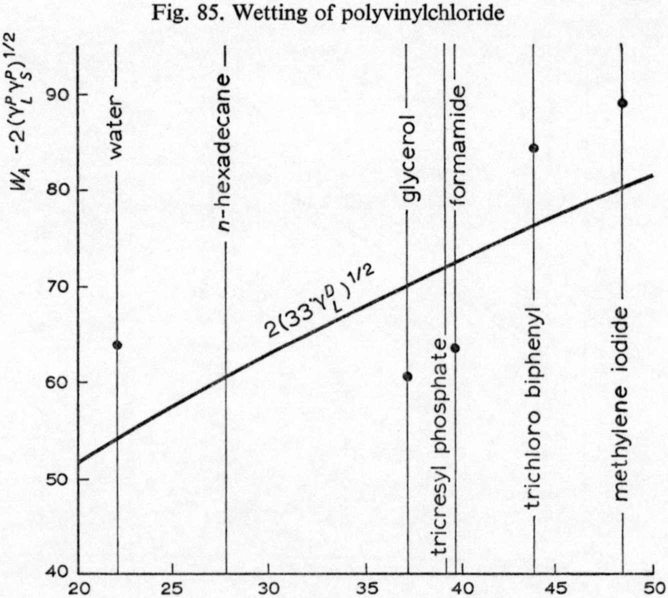

Fig. 86. Wetting of polystyrene

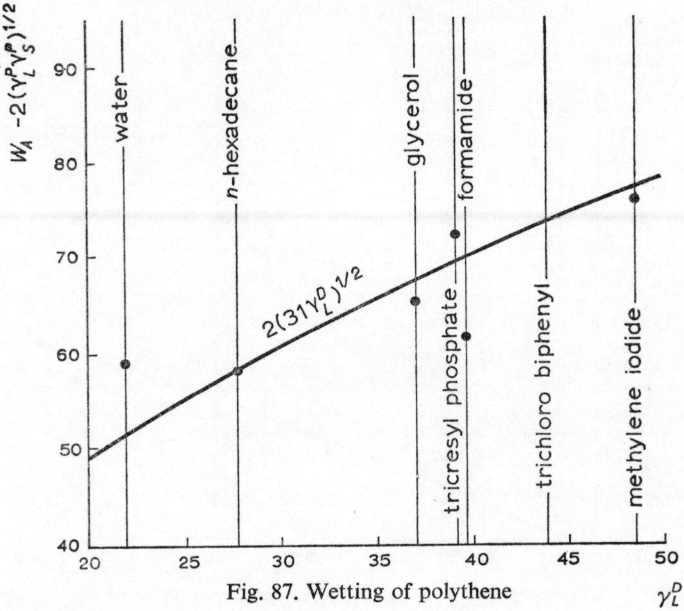

Fig. 87. Wetting of polythene

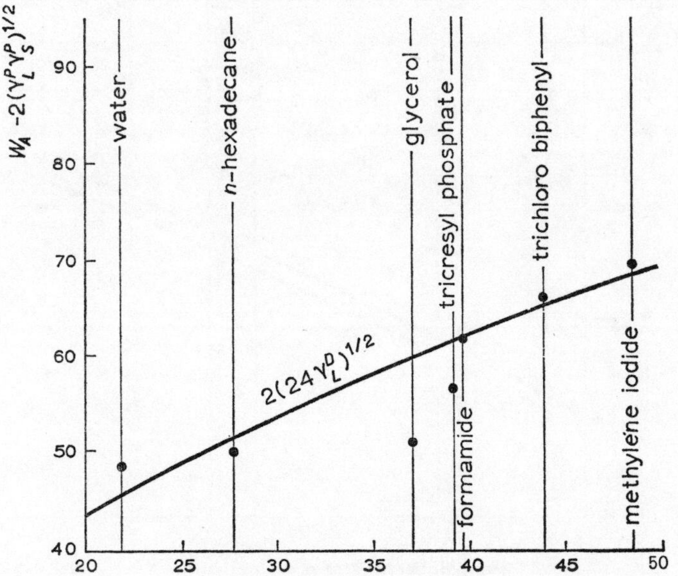

Fig. 88. Wetting of octadecylamine monolayer

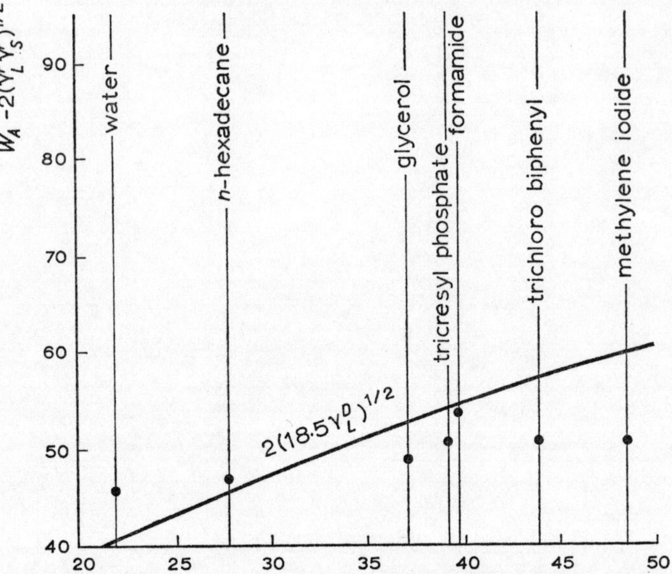

Fig. 89. Wetting of polytetrafluoroethylene γ_L^D

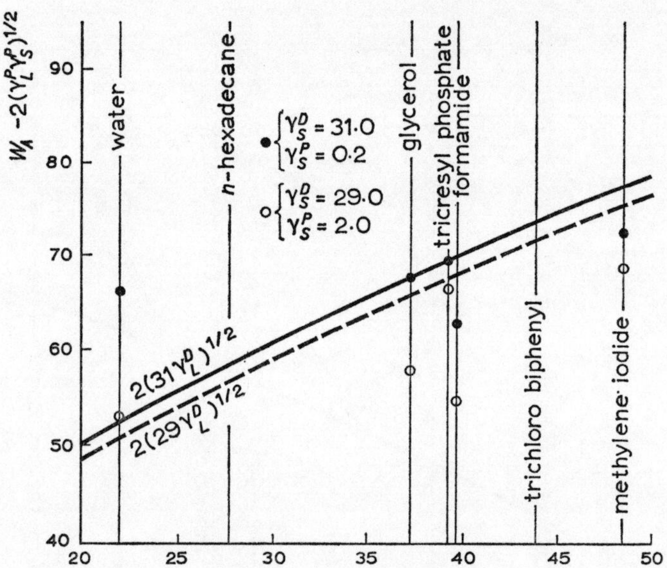

Fig. 90. Wetting of polychlorotrifluoroethylene γ_L^D

and $2(\gamma_c \gamma_L^D)$ where $W_a = \gamma_L(1 + \cos \theta)$. The results are shown in Figs. 84 to 90. $2(\gamma_c \gamma_L^D)^{\frac{1}{2}}$ is shown as the solid curve drawn through the data.

Although there is considerable scatter of data, the quantity $[W_a - 2(\gamma_L^P \gamma_S^P)^{\frac{1}{2}}]$ generally shows that same variation with γ_L^D as does $2(\gamma_c \gamma_L^D)^{\frac{1}{2}}$. Polystyrene shows the greatest scatter of data, and polytetrafluoroethylene and polychlorotrifluoroethylene show lesser slopes than predicted. Generally the γ_S^P calculated from the glycerol or dimethylformamide data is too low when applied to contact angle data for water. Fig. 90 shows the result of (1) adjusting γ_s^P to bring the glycerol and dimethylformamide data into coincidence with $2(\gamma_c \gamma_L^D)^{\frac{1}{2}}$ and (2) adjusting both γ_c and γ_s^P to bring the results for water into line. These shifts could probably be rationalized by detailed considerations of liquid-solid interactions such as hydrogen bonding and group spacings, but that is beyond the scope of this paper. It does appear that the term $2(\gamma_c \gamma_L^D)^{\frac{1}{2}}$ gives a fair approximation of the dispersion force contribution to the work of adhesion and that the estimated polar force contribution, $2(\gamma_S^P \gamma_L^P)^{\frac{1}{2}}$, when added to $2(\gamma_S^D \gamma_L^D)^{\frac{1}{2}}$ gives a work of adhesion in reasonably good agreement with the measured work of adhesion of liquids of polarity intermediate between water and hydrocarbons. The works of adhesion between the solids and the pressure sensitive adhesive were calculated from

$$W_a = 2\,[(\gamma_A^D \gamma_S^D)^{\frac{1}{2}} + (\gamma_A^P \gamma_S^P)^{\frac{1}{2}}] \tag{5}$$

where γ_A^D and γ_A^P now represent the dispersion and polar force contributions to the surface energy of the pressure sensitive adhesive. γ_A^D and γ_A^P were also estimated from contact angle data.

The measurement of contact angles of liquids on such soft viscoelastic polymers as pressure sensitive adhesives is fraught with difficulties. Solubility of the adhesive or adhesive components in the test liquid and the distortion of the adhesive surface by the liquid are serious problems. The value of γ_S^D and γ_S^P of 16 and 4 erg per cm^2, respectively, for the adhesive used in these experiments are at best quite approximate,* but they are in general agreement with values obtained on several pressure sensitive adhesives. Apparently, pressure sensitive adhesives have rather low surface energies.

Table 27 tabulates the γ_S^D and γ_S^P values of the solids and the

*Calculated by the Kaelble and Uy equation from θ(water)=95·2° and θ(methylene iodide)=77·4°.

works of adhesion between adhesive and solid calculated by Equation (5). Re-evaluated surface energies of some cellulose esters described in a previous publication[9] are also included. The γ_S^D and γ_S^P values for cellulose triacetate were calculated from water and methylene iodide contact angle data. The values for cellulose tricaprylate, cellulose trilaurate, and cellulose tristearate, are, however, only estimates which provide a solution to the following equation,

$$72{\cdot}8(1 + \cos\theta) = 2(\sqrt{21{\cdot}8\,\gamma_S^D} + \sqrt{51{\cdot}0\,\gamma_S^P})$$

where θ is the contact angle of water on these surfaces. It is obvious that the earlier estimates of the surface energies of these cellulose esters were generally too high, though the γ_S^D values of the higher esters remain the same as the previously estimated values of γ_c.

Table 28 also lists the boundary cleavage stress calculated from peel force data. In Fig. 91 the work of adhesion is plotted against

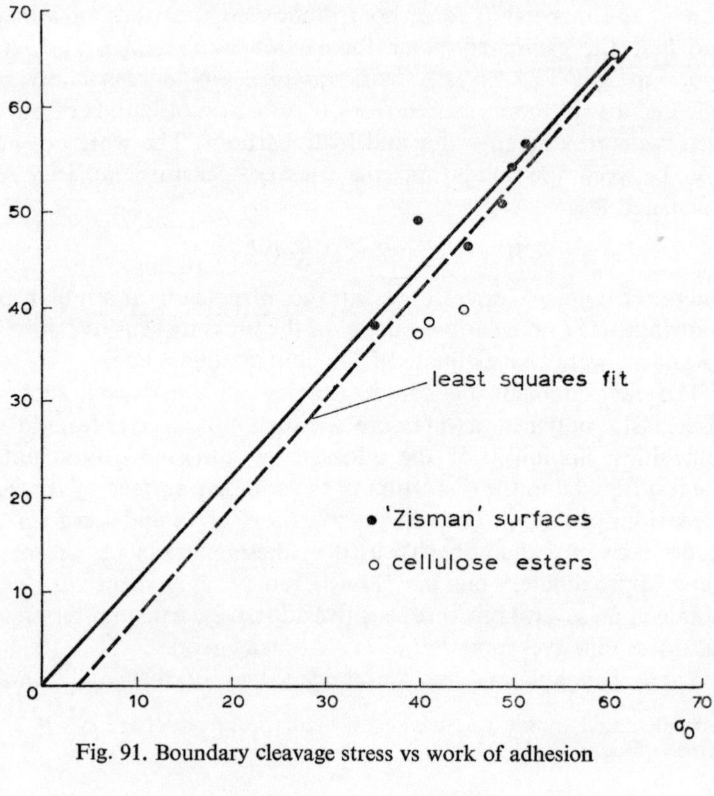

Fig. 91. Boundary cleavage stress vs work of adhesion

the boundary cleavage stress on rectangular coordinates and the relationship appears to be approximately linear. Polyethylene falls rather badly out of line. This can be explained in part by an adhesion retention of only about 80 per cent in the adhesive after removal from the polyethylene, indicating the existence of a weak boundary layer which transferred to the adhesive.

Discussion and conclusions

Analysing these results, we see a direct relationship between the work of adhesion and the boundary cleavage stress. We cannot rationalise this on the basis that separation is by cohesive failure in the adhesive because

1. The cleavage stress is much less than the ultimate tensile strength of the adhesive, and
2. the cleavage stress should not vary if failure occurred in the adhesive.

We cannot argue that cohesive failure occurs in the adherend because even the weakest of these, the cellulose tristearate, has a tensile strength of 50 kg per cm², and this is higher than the boundary cleavage stress. We can, of course, rationalise the results on the basis of weak boundary layers on the solids or weak boundary layers produced by solid-adhesive interaction, but it seems unlikely that these should then display a regular progression of cohesive strengths. Furthermore, except in the case of polyethylene, there was no deterioration in the peel adhesion strength of the tape to indicate pick-off of a weak boundary layer.

We can argue that the change in the boundary cleavage stress is only a matter of less wetting of the low energy surfaces and more complete wetting of the high energy surfaces. However, if we apply de Bruyne's criteria of wetting we must conclude that an adhesive so low in surface energy should wet all of these surfaces.

If we apply Huntsberger's criterion for interfacial failure, $\phi(\gamma_S/\gamma_A)^{\frac{1}{2}} < 1$, obtaining ϕ from the relationship

$$2\phi(\gamma_S\gamma_A)^{\frac{1}{2}} = 2[(\gamma_{SD}\,\gamma_{AD})^{\frac{1}{2}} + (\gamma_S^P\,\gamma_A^P)^{\frac{1}{2}}]\,,$$

we see that interfacial failure is indicated in the separation of the adhesive from polytetrafluoroethylene, cellulose tristearate, and cellulose trilaurate, but not from the other surfaces. If the criterion applies, we should expect to see two sets of cleavage stresses, one at a low level where interfacial separation occurs, the other at a higher level, where cohesive failure occurs in the adhesive. This is not evident in the data.

In view of the above it seems that we must revise or extend our views on the mechanism of cleavage of the pressure sensitive adhesive bond. The idealised bond visualised in the calculation of forces across the interface takes into account mismatches in molar volumes and in distribution of energy sites, but it does not take into account the co-operative interaction of chain atoms in long polymer molecules. An interface which consists of long, entangled molecules in the rubbery state adsorbed on a surface of long entangled molecules in the glassy state, and in which there is no interentanglement of molecules across the interface, will be weaker than the weakest of the two solids forming the bond. An interchain link which cannot slip or does not have time to slip is effectively a covalent bond, and this is many fold stronger than a van der Waals' bond.

The probabilities of failure of bonds under stress can be compared in terms of the activation energies for dissociation of these bonds, typically about 80 kcal per mol for the covalent bond and 5 to 10 kcal per mol for the van der Waals' bond. If we apply the Arrhenius expression to the fracture process, the relative probabilities of fracture under stress will be

$$\frac{P\text{ van der Waals'}}{P\text{ covalent}} = \frac{\exp\dfrac{80{,}000}{RT}}{\exp\dfrac{10{,}000}{RT}} = \exp\frac{70{,}000}{RT} = 10^{50}$$

Thus any fracture process that proceeds in stepwise fashion along a polymer chain will in all probability result only in fracture of van der Waals' bonds. If the interface is the one unique zone in which there are only van der Waals' bonds, it is highly probable that separation will occur there. Even though van der Waals' bonding in the bulk of the adhesive is weaker than the van der Waals' bonding across the interface, the presence of entanglements tips the scale very much in favour of the propagation of failure along the interface.

The interfacial attraction will, of course, determine the energy which must be supplied to rupture the bonds and this will be reflected in the total mechanical work and in the peel force, hence the correlation between the work of adhesion and the cleavage stress.

TABLE 26

Dispersion and Polar Contributions to Liquid Surface Energy (T = 20°C)[15]

Number	Liquid	γ_L (dyne/cm)	γ_L^D (dyne/cm)	γ_L^P (dyne/cm)
1	Water	72·8	21·8	51·0
2	Glycerol	63·4	37·0	26·4
3	Formamide	58·2	39·5	18·7
4	Methylene iodide	50·8	48·5	2·3
5	Trichlorobiphenyl (Aroclor 1242)	45·3	44·0	1·3
6	Tricresylphosphate	40·9	39·2	1·7
7	n-Hexadecane	27·6	27·6	0·0

TABLE 27

Dispersion and Polar Contributions to Solid Surface Energy

Surface	γ_S^D	γ_S^P
1. Polyethylene terephthalate	43·0	1·0
2. Polyvinyl chloride	39·0	1·4
3. Polystyrene	33·0	1·6
4. Polyethylene	31·0	1·4
5. Octadecylamine monolayer	24·0	0·43
6. Polytetrafluoroethylene	18·5	0·11

TABLE 27(a)

Dispersion Force Interactions of Liquids of Table 27 and Solids of Table 28

Surface liquid	1	2	3	4	5	6
	$\gamma_L(1 + \cos\theta) - 2\,\gamma_S^P\gamma_L^P$					
Water	69·8	59·7	63·5	50·8	48·3	45·6
n-Hexadecane	—	—	—	—	50·0	46·8
Glycerol	73·8	76·0	60·2	63·3	51·0	49·0
Tricresylphosphate	—	77·5	—	71·7	56·5	50·6
Formamide	75·6	70·6	63·3	61·1	61·6	53·3
Trichlorobiphenyl	85·3	—	84·6	—	65·8	50·7
Methylene iodide	89·4	88·3	88·6	78·5	69·5	50·6

TABLE 28

Cleavage stress in the Pressure Sensitive Adhesive Bond and work of Adhesion.

Surface	γ_S^D (erg per cm²)	γ_S^P (erg per cm²)	Peel Force (a)	σ_o (b)	Wa $2(\gamma_S^D\gamma_A^D)^{\frac{1}{2}}+2(\gamma_S^P\gamma_A^P)^{\frac{1}{2}}$ (c)
Cellulose triacetate	29	12·5	850	60·3	66·8
Polyethylene terphthalate	43	1·0	610	51·0	57·5
Polyvinyl chloride	39	1·4	580	49·8	54·7
Polystyrene	33	1·6	550	48·5	50·8
Polychloro tri fluoroethylene	31	0·2	470	44·8	46·3
Cellulose tricaprylate	21	0·6	455	44·1	39·8
Cellulose trilaurate	20	0·5	390	40·8	38·6
Polyethylene	31	1·4	365	39·5	49·3
Cellulose tristerate	19	0·4	365	39·5	37·4
Polytetra fluoroethylene	18·5	0·7	290	35·2	37·7

(a) Grams per 1·27 cm at 100 cm per minute after 1 week of contact under 16 g per cm².

(b) Calculated from $P=\dfrac{\sigma_o^2 K^2 a\, b}{2\gamma\,(1\text{-}\cos 90°)}$

a = 0·0025 cm, b = 1·27 cm, γ = 6·8×10³ g per cm², $K \cong$ 1·0, P is in kg per cm².

(c) γ_A^D = 16·0 erg per cm²; γ_A^P = 4·0 erg per cm².

Acknowledgement

The author expresses his appreciation to D. H. Kaelble for advice and stimulating discussion, to D. R. Beresford for contact angle measurements on the pressure sensitive adhesive, and to J. O. Hendricks for advice and encouragement.

References

1. Bikerman, J. J., Making Polyethylene Adhesionable, *Adhesives Age*, **2**, (2), 23, 1959.
2. Sharpe, L. H. and Schonhorn, H., Contact Angle, Wettability and Adhesion, *Advances in Chemistry Series*, Ed. Gould, R. F., Am. Chem. Soc., (43), 1964.

3. Schonhorn, H. and Hansen, R. H., Surface Treatment of Polymers for Adhesive Bonding, *J. Appl. Polymer Sci.*, **11**, 1461, 1967.
4. Gilman, J. J., Direct Measurements of the Surface Energies of Crystals, *J. Appl. Phys.*, **31**, 2208, 1960.
5. Lifschitz, E. M., Attraction between Non-metallic Solids, *Dokl. Akad. Nauk SSSR*, **97**, 643, 1954.
6. Gardon, J. L., Variables and Interpretation of Some Destructive Cohesion and Adhesion Tests, *Treatise on Adhesion and Adhesives*, Ed. Patrick, R. L., Vol. 1, Marcel Dekker, New York, 1967.
7. Good, R. J., Intermolecular and Interatomic Forces, Ibid.
8. Huntsberger, J. R., The Mechanisms of Adhesion, Ibid.
9. Dahlquist, C. A., Adhesion and Cleavage Stress at Interfaces Between Solids and a Pressure Sensitive Adhesive, Special Technical Publication (360), Am. Soc. for Testing and Materials, 1963.
10. Kaelble, D. H., Theory and Analysis of Peel Adhesion: Mechanisms and Mechanics, *Trans. Soc. Rheology*, **3**, 161, 1959.
11. Kaelble, D. H., Theory and Analysis of Peel Adhesion: Bond Stresses and Distributions, Ibid, **4**, 45, 1960.
12. Kaelble, D. H., Peel Adhesion: Micro-fracture Mechanics of Interfacial Unbonding of Polymers, Ibid, **9**, Part 2, 135, 1965.
13. De Bruyne, N. A., The Extent of Contact Between Glue and Adherend, Aero Research Tech. Notes, Bull. (168), Dec., 1956.
14. Zisman, W. A., Relation of the Equilibrium Contact Angle to Liquid and Solid Constitution, *Advances in Chemistry Series*, Ed. Gould, R. F., Am. Chem. Soc., (43), 1964.
15. Fowkes, F. M., Dispersion Force Contributions to Surface and Interfacial Tensions, Contact Angles, and Heats of Immersion, Ibid.
16. Uy, K. C. and Chang, R., Approaches to Achieve Adhesion, Adhesive Restorative Dental Materials II, Public Health Service Publication (1494), U.S. Govt. Printing Office, Washington, D.C., 1966.
17. Kaelble, D. H. and Uy, K. C., A Reinterpretation of de Bruyne's Adhesion Rule. To be published.
18. Hendricks, J. O. and Dahlquist, C. A., Pressure Sensitive Adhesive Tapes, *Adhesion and Adhesives*, Eds. Houwink, R. and Salomon, G., Vol. 2, Elsevier, New York, 1967.
19. Zisman, W. A., Constitutive Effects on Adhesion and Adhesion, *Adhesion and Cohesion*, Ed. Weiss, P., Elsevier, New York, 1962.

The Nature of Adhesive Forces

C. WEAVER, B.Sc., D.Sc., F.Inst.P., M.I.E.E., M.I.E.E.E.
THE UNIVERSITY OF STRATHCLYDE, GLASGOW

A discussion of the strength of interfaces. The experimental strength of most joints can be explained in terms of van der Waals forces alone, but there is an increasing amount of evidence to show that the strength of an interface may be considerably greater than the ultimate strength of the component materials. Under such conditions, failure under tension occurs in regions remote from the interface. If these conditions can be achieved in an adhesive joint, the strength of the joint depends on the strength of the individual components, where the adhesive layer is regarded as a separate component, and the thickness of the adhesive can be an important factor.

Some instances of high adhesion are discussed to show the nature of the factors involved.

Introductory

Adhesion is frequently discussed in relation to the strength of joints as determined experimentally; but a joint is a relatively complex system where the stresses are unlikely to be uniform and when failure occurs it may be in the layer of adhesive, in either component of the joint, or at either of the two interfaces between the adhesive layer and the joint components. The term adhesion should be reserved strictly for the bonding at the interfaces since the other three possible points of joint failure depend on the cohesive strengths of the materials involved. If we assume that the interfaces represent the weakest points so that failure occurs here under stress, the empirical strength of most joints can be explained in terms of van der Waals' forces alone, acting at the interface. This was pointed out long ago by de Bruyne,[1] and shows that relatively weak bonding at interfaces can explain experimental joint strengths, provided that there is intimate contact over the entire interface.

Attempts to use joint strengths as a measure of adhesion can be extremely misleading because of the assumptions involved. In the first place, it is difficult to calculate the stress acting at an interface and even more difficult to avoid stress concentrations which would produce progressive failure, although considerable work has been done on the calculation and experimental determination of stress distributions.[2] It is also dangerous to assume intimate contact over an entire interface, even when an adhesive is applied in liquid form to a solid surface, a point which has been discussed in considerable detail by Sharp and Schonhorn[3] and forms the essential basis of their theories. The final point is whether the interfaces may be regarded as weak in comparison to the ultimate strengths of the adhesive and/or the joint components, and the present discussion is intended to cover only the nature and magnitude of the bonding at interfaces as compared with the ultimate strengths of materials.

Since adhesion involves intimate contact of two surfaces, it is often convenient to think in terms of the energies of the surfaces involved and Zisman[4] introduced the distinction between high energy surfaces and low energy surfaces. Most liquids have surface free energies below 100 erg/cm^2 at room temperature; solid surfaces having similar free energies are termed low energy surfaces, whereas hard solids having surface free energies above about 500 erg/cm^2 are termed high energy surfaces. Organic polymers, waxes and similar materials generally have low surface energies whilst metals, metal oxides and glasses form high energy surfaces. Materials which have low energy surfaces generally have a molecular structure held together by van der Waals' forces and the adhesion between two such surfaces will tend to be of the same type. The work required to separate the two surfaces may be estimated, using conventional ideas of surface tension or surface energy. If a layer of organic adhesive which has been allowed to set on a solid surface may be regarded as equivalent to a liquid, the force required to strip it from the surface may be estimated as of the order of 6×10^8 dyne/cm^2[5,6] and failure will generally occur within the adhesive rather than at the interface.

The dispersion effect generally forms the main contribution to the van der Waals' forces and consequently the strength of the bonding will depend on the polarisabilities of the materials involved, but it would appear that the ultimate strength of interfaces under ideal conditions may be of the order of $10^7 - 10^9$ dyne/cm^2 without considering any other bonding mechanism. This is comparable with the ultimate strengths of many materials and higher than the experimental strengths of most joints.

Measurement of adhesion

It is difficult to obtain reliable and unambiguous measures of adhesion. It is almost impossible to obtain perfectly smooth solid surfaces. Surface irregularities and minute protuberances prevent perfect contact between two solid surfaces. Contact occurs only at the high spots and if pressure is applied to obtain better contact, the elastic stresses in the deformed asperities cause a partial separation of the surfaces as soon as the pressure is removed. The effective area of contact is uncertain and difficult to determine, so that the force required to separate the surfaces gives no clue to the true strength of the bonding at the points where contact does occur. The effects may be overcome to some extent by using a soft, ductile material for one surface and experiments have been described by Bowden and Tabor[7] where a ball-bearing or similar indenter was pressed into a lead or indium surface. Plastic flow occurs under pressure, and the force required to separate the two surfaces becomes equal to the compressive force originally applied. Similar results may be obtained by annealing two metals in contact under pressure so as to reduce the elastic stresses. In all these experiments, fragments of one metal frequently remain attached to the other surface, suggesting that the interface may be stronger than the materials used provided the surfaces are absolutely clean.

These methods are, of course, equivalent to using one surface which is initially liquid since plastic flow occurs. Intimate contact may be achieved by freezing a liquid *in situ* on a solid surface and this method has been used to investigate the adhesion of ice[8]. For ice on metal surfaces, shearing forces up to 40 kg/cm^2 (4×10^7 dyne/cm^2) were required to break the bonding and failure invariably occurred in the ice rather than at the interface. This figure is probably still low since corresponding friction experiments with ice gave shear strengths up to 120 kg/cm^2 ($1{\cdot}2 \times 10^8$ dyne/cm^2). Considerably lower figures were obtained for ice on plastic surfaces[6] but the contact angles for water on these materials were always high, and this could result in an imperfect interface according to the ideas of Sharp and Schonhorn.

Some of the difficulties of obtaining perfect contact between two solid materials may be eliminated by depositing a film on a clean surface under high vacuum conditions. Such a film faithfully contours all surface irregularities and almost perfect contact between film and substrate is readily obtained. The major difficulty is to remove the film in a manner which gives a quantitative measure of the forces involved. Various methods have been tried but one of the most reliable methods[9] used a film deposited on the rotor of an ultra-centrifuge. The rotor was spun at high speed until the film was detached and a figure of $1 \cdot 1 \times 10^9$ dyne/cm^2 was obtained for the adhesion of a poorly adherent film of chromium. For most metal films the rotor tended to burst before the film was detached and this particular value was for a film which had been deposited on an oiled surface. The figure may therefore be regarded as representing van der Waals' bonding and is probably close to the practical limit of this method of measurement. Other methods generally tend to be even more restricted or to give anomalous results.

These considerations led to the detailed investigation and development of a method of measuring film adhesion, in which a smoothly-rounded chrome-steel point is drawn across the coated surface. The load on the point is gradually increased until a critical load is reached at which the film is removed by the point leaving a clear channel.[10] It was shown that the critical load varied with the nature of both film and substrate but could not be related directly to the mechanical properties of either, and was also independent of film thickness except for very thin films <300 Å where incomplete coverage of the substrate might be expected because of aggregation during the early stages of film growth. This was followed by a detailed analysis which showed that a shearing force is produced between film and substrate as the substrate is deformed by the applied load. The original assumption that the measurement was determined entirely by the interface has recently been confirmed in a striking manner by measurements on double-layer metal films on glass. The results shown in Fig. 92 were obtained for different thicknesses of gold deposited on glass and then overlaid with aluminium. Gold has a relatively poor adhesion to glass and the measurements start at a low value. Aluminium has a much higher adhesion to glass and there is an abrupt increase in the measured adhesion of the two-layer film when the aluminium diffuses right through the gold layer and reaches the metal-glass interface. This may readily be detected by observation of the interface through the glass to see the change in reflectivity. The time lapse before the increase occurs is proportional to the square of the thickness of the gold layer and the effect is readily seen in the diagram.

The analysis of this method showed that the shearing force at the

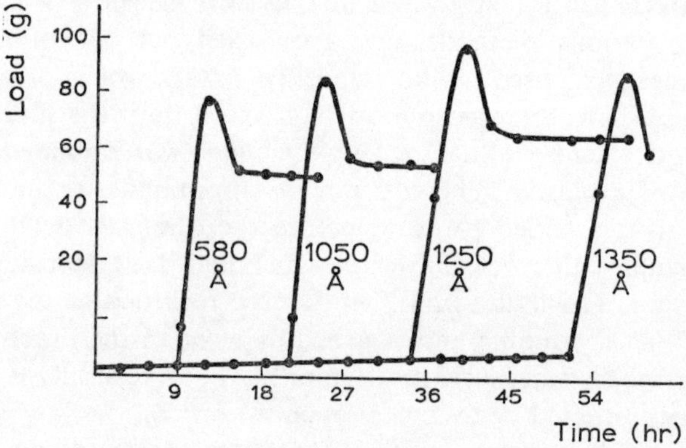

Fig. 92. Adhesion of 2-layer films of aluminium on gold, deposited on glass at room temperature and aged at 50°C

film-substrate interface could be calculated from the critical load on the point, and for poorly adherent films the shearing force was of the order of 10^9 dyne/cm^2, which is in good agreement with the figures obtained by other methods. Because of the necessity of deciding the exact load at which a clear channel is produced, the method is particularly suitable for metal films on transparent substrates. It has been used extensively for such cases and under these conditions the results are reproducible to ± 5 per cent. Another restriction is imposed because the substrate must deform to produce the required shearing force and consequently it is difficult to obtain any consistent results with extremely hard substrates such as sapphire. Such substrates are harder than the metal point so that the point deforms. Some work has been done using diamond styli but few results of value have been obtained.

The restrictions on the method have been emphasised but on the other hand, there are some remarkable advantages. Adhesion measurements have been made over a range from below 10^7 dyne/cm^2 to 3×10^{10} dyne/cm^2. The lower limit could easily be extended by using a point with a larger tip radius but the upper limit is imposed by the strength of the substrate and glass or silica tends to splinter or crack at higher loads. This upper limit is however much higher than can be achieved by other methods and it has allowed adhesion to be measured in an otherwise inaccessible range. Most methods are limited because the ultimate strength of many materials in tension is of the order of 10^9 dyne/cm^2 and consequently failure occurs away from the interface, as in the case of ice on metals mentioned above. The extended range is obtained because the method operates in compression and compressive strengths tend to be higher than tensile strengths for many materials. This means that it is possible to obtain a measure of adhesive bonding which exceeds the ultimate tensile strengths of the components. The other major advantage is the ease of obtaining an interface with perfect contact and freedom from contamination.

Metal films on glass

Measurements of the adhesion of metal films deposited on glass or silica have shown that for a non-reactive metal like gold, the adhesion remains constant at a comparatively low value of the order of 10^9 dyne/cm^2, a figure which can be completely explained in terms of van der Waals' forces. However, measurements on other metals frequently show an increase in adhesion with time and in many cases the adhesion rises to 3×10^{10} dyne/cm^2, or possibly higher since this limit is imposed by the strength of the substrate.[11] Fig. 93 shows a

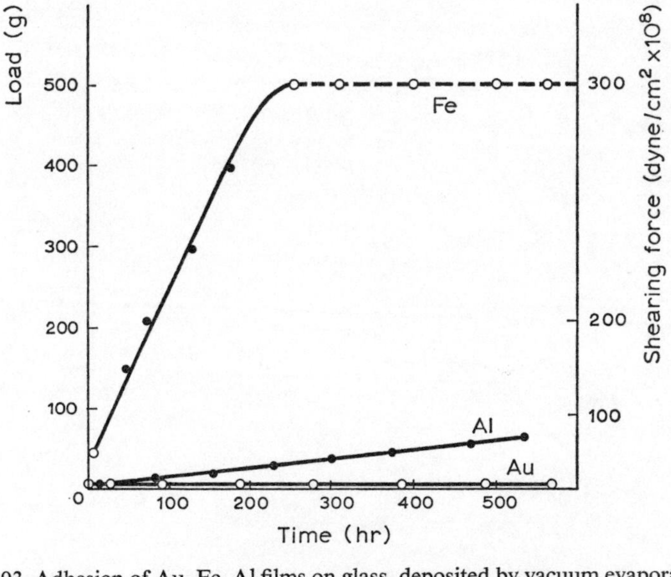

Fig. 93. Adhesion of Au, Fe, Al films on glass, deposited by vacuum evaporation and aged at atmospheric pressure

direct comparison of results obtained for Au, Al, Fe deposited under conventional vacuum conditions (10^{-5} mm Hg) on glass and subsequently measured at atmospheric pressure. The adhesion of the iron rose steadily to a value almost two orders of magnitude greater than the adhesion of gold, and, from the shape of the curve, might have risen higher if measurements could have been continued above the final level shown but the glass substrate tended to crack or splinter at higher loads.

The final adhesion was much higher than could be attributed to intermolecular forces and it was suspected that the effects were due to oxide formation. This is confirmed by the results for iron films shown in Fig. 94. The films were deposited both at a higher residual air pressure and also after the vacuum system had been flushed out with hydrogen and finally brought to the same pressure (10^{-3} mm Hg). The higher oxygen concentration produced a higher initial adhesion and more rapid ageing while the specimens prepared in hydrogen had a much lower adhesion over the entire ageing period. Similar curves for aluminium are shown in Fig. 95, and films prepared at the higher oxygen concentration could not be removed without splintering the glass even immediately after deposition, whereas films pre-

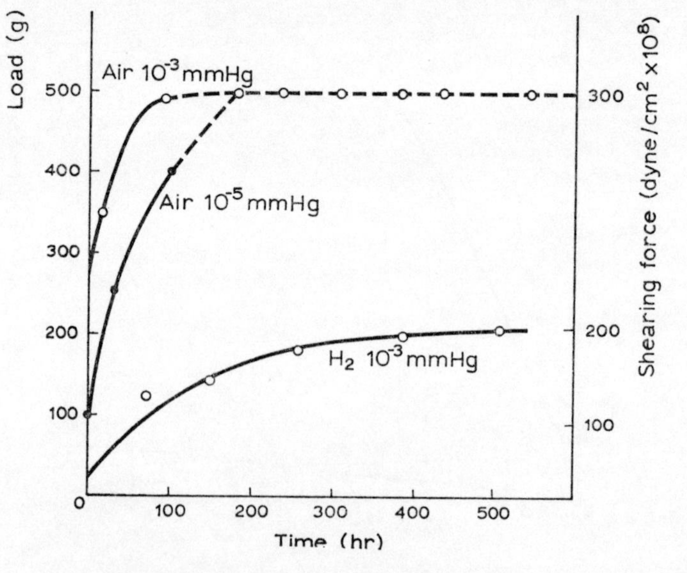

Fig. 94. Effect of residual gas upon the adhesion of iron films vacuum-deposited on glass and aged at atmospheric pressure.

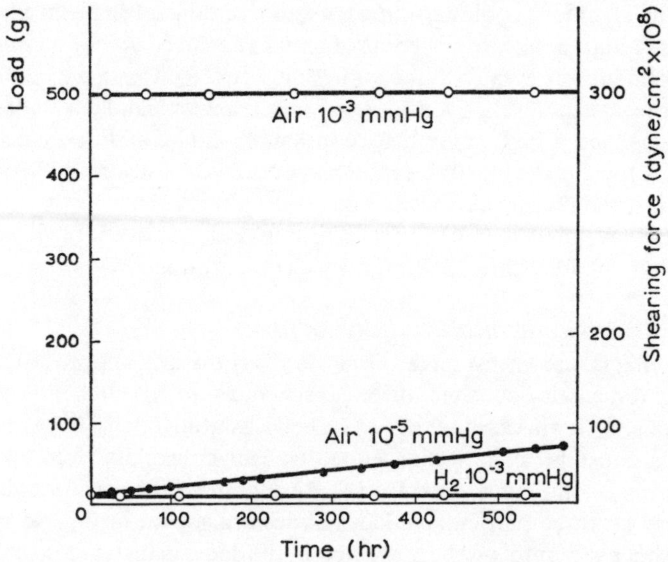

Fig. 95. Effect of residual gas upon the adhesion of aluminium film vacuum-deposited on glass and aged at atmospheric pressure

pared in a residual hydrogen atmosphere showed only van der Waals' bonding.

These results all indicate a reaction to form oxide at the interface and produce a form of chemical bonding. The oxygen may be trapped in the film during deposition or may diffuse through the film to the interface after deposition. Aluminium shows somewhat peculiar behaviour as compared to iron because it produces continuous films which are relatively free of major structural defects such as grain boundaries and when exposed to the atmosphere it forms a protective oxide skin. The adhesion of aluminium is therefore affected only by oxygen which is trapped during deposition, although the effects can be very pronounced when sufficient oxygen is present because of the strong affinity of aluminium for oxygen. The results of Fig. 95 show immediate formation of a very strong bond when sufficient oxygen is available during deposition. On the other hand, when oxygen is excluded during deposition, the adhesion remains at a low value corresponding to van der Waals' bonding because a protective oxide skin is formed on the free outer surface which effectively excludes atmospheric oxygen from the interface.

One significant point about the results obtained for metal films on glass is that in all cases where the changes occurred slowly enough for convenient observation, the adhesion started at a low value and increased steadily. In a few cases of very reactive metals, the adhesion had reached a high value before measurements could be made, but, if we allow for the inevitable interval between deposition and measurement, it would appear that the adhesion always starts at the van der Waals' level and then changes. This might alternatively be described as physical adsorption followed by chemisorption.

Metals on alkali halide surfaces

The effects shown by metal films on glass do not appear when the same films are deposited on cleavage faces of alkali halide single crystals.[12] In all cases which have been examined, the adhesion remains constant at its initial value and can be explained completely in terms of van der Waals' forces. Results obtained for a number of metals on three particular alkali halides are shown in Fig. 96 where the shear strength of the interface is plotted against the calculated van der Waals' energy. The line shown does not pass through the origin because it has been assumed that a reasonable estimate of the van der Waals' energy can be obtained by considering the dispersion effect alone. If this line is produced to meet the horizontal axis, there is a small negative intercept corresponding to the minor interactions which have been neglected.

These results do not have any practical significance but they are particularly interesting because they show clearly the strength which can be obtained from van der Waals' bonding alone, and also that the strength of the bonding can easily vary by an order of magnitude or more according to the materials involved. Furthermore, they show that even with oxidisable metals, there will be formation of oxide at the interface only when the conditions are favourable. At a metal-glass interface, an oxide layer will tend to give a continuous transition from the silica network of the glass to the metal, forming a strong bond to each. There is an implication that such a structure must have a lower energy than the original glass-metal interface.

Other materials

The available information on the adhesion of metal films to other materials is not so complete. Measurements of the adhesion of metal films to polymer surfaces have shown that the adhesion frequently

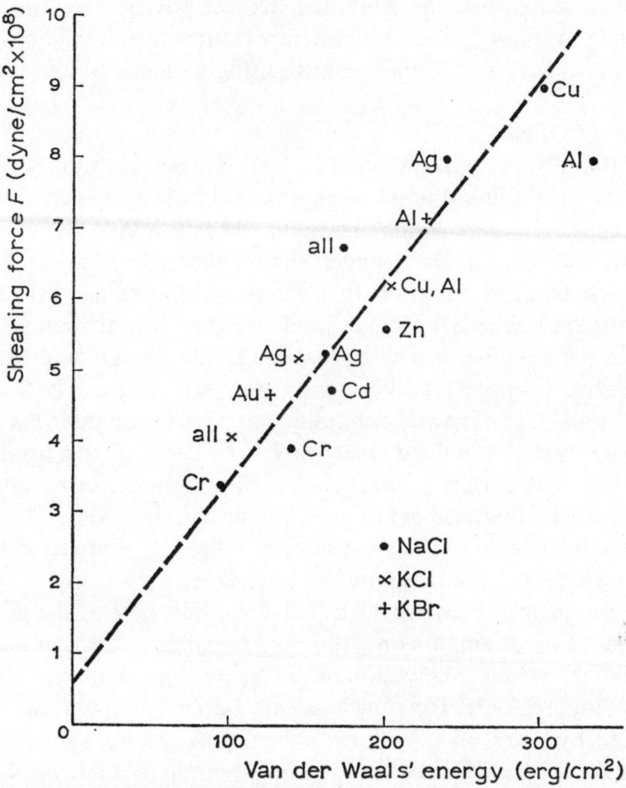

Fig. 96. Relation between measured adhesion of metal films on alkali halide (100) faces and the calculated Van der Waals' energies due to dispersion forces

increases with time, reaching values which are significantly greater than may be attributed to van der Waals' forces alone. There is however, no relationship with the pattern which has already been established for metal films on glass and there can be no question of an oxide film playing a similar role in this particular case. Instead, there is evidence of an electrical component of adhesion due to a charge transfer mechanism which produces an electrical double layer at the interface. Work in this field is still in progress and results are not yet complete.

There is also evidence of high adhesion values obtained with bulk materials and one notable case occurs when a droplet of molten aluminium is allowed to solidify on a sapphire surface. It is almost impossible to remove the solidified droplet without fracturing the underlying sapphire. At such high temperatures there could of course be some diffusion and it may be misleading to speak of an interface.

Conclusions

One of the most important points which has been established is the possibility of obtaining bond strengths at interfaces which considerably exceed van der Waals' bonding. The interface between two materials may be much stronger than either component. At first sight, such bond strengths may seem to be unnecessary since they would not lead directly to higher joint strengths but, if such values of adhesion are possible, it may be feasible to obtain adequate strength with relatively imperfect interfaces. This term is used here in the sense adopted by Sharp and Schonhorn to describe an interface where perfect contact is obtained over only a fraction of the total area. Alternatively, if perfect contact and high adhesion can be obtained simultaneously, the problem of joint strength is effectively transferred from the interfaces to the component materials, which would of course include any intermediate adhesive layer.

It has frequently been advocated that the thickness of the adhesive layer should be as small as possible for maximum strength and it is clear that the smaller the volume of adhesive, the smaller the chance of including major defects such as air bubbles. This view is also supported by work on thin fibres, where greatly increased strengths have been observed. The increase in strength in such cases is of course largely due to the elimination of surface defects or Griffiths cracks and it is uncertain whether such defects can be eliminated when a layer of adhesive is sandwiched between two joint members. There is little information to show how thin an adhesive layer must be to obtain increased strength. There are some measurements on thin metal films where a pronounced increase in ultimate strength has been observed for film thickness below 2 500 Å.[13] Other workers have found strengths which are significantly higher than bulk material for films up to several microns thick.[14] These effects have been attributed in part to the pinning of dislocations at the boundary surfaces and similar effects may occur in adhesive films even though it is hardly reasonable to assume that the results for metal films can be used to predict the behaviour of adhesive films. If increased film strength can only be obtained with films of the order of 2 500 Å, it would imply working with joint surfaces which are optically flat before such thicknesses of adhesive would be possible. This is clearly impracticable but so are the grounds for assuming this figure and further work in this field seems to be necessary.

References

1. De Bruyne, N. A., *J. Sci. Instr.*, **24,** 29, 1947.
2. Eley, D. D., *Adhesion*, Oxford University Press, 1961.
3. Sharpe, L. H. and Schonhorn, H., *Advances in Chemistry Series*, **48,** 189, 1964.
4. Fox, H. W. and Zisman, W. A., *J. Colloid Sci.*, **5,** 514, 1950.
5. Tabor, D., *Report Progr. Appl. Chem.*, **36,** 621, 1951.
6. Bowden, F. P. and Tabor, D., *Friction and Lubrication of Solids*, Vol. II, Oxford University Press, 1964.
7. Bowden, F. P. and Tabor, D., *Friction and Lubrication of Solids*, Vol. I, Oxford University Press, 1957.
8. Raraty, L. E. and Tabor, D., *Proc. Phys. Soc.*, **B68,** 625 1958.
9. Beams, J. W., *43rd Ann. Proc. Amer. Electroplaters Soc.*, 1956.
10. Benjamin, P. and Weaver, C., *Proc. Roy. Soc.*, **A254,** 163, 1960.
11. Benjamin, P. and Weaver, C., *Proc. Roy. Soc.*, **A254,** 177, 1960.
12. Benjamin, P. and Weaver, C., *Proc. Roy. Soc.*, 1963.
13. Beams, J. W., Breazeale, J. B. and Bart, W. L., *Phys. Rev.*, **100,** 1657, 1955.
14. Menter, J. W. and Pashley, D. W., *Structure and Properties of Thin Films*, p. 111, Wiley, New York, 1959.

Glow Discharge Pretreatment of Polyethylene

B. W. MALPASS, B.Sc., Ph.D., and K. BRIGHT, B.Sc., Ph.D.**

PASSFIELD RESEARCH LABORATORIES

Introductory

Polyethylene is a low surface energy material and therefore is wetted only with difficulty by polar liquids. Accordingly, it is necessary, when polyethylene is to be printed upon or adhesively bonded, to convert its surface to one of relatively high energy. The work described here is concerned with the use of adhesives of high cohesive strength to bond polyethylene to polyethylene in such a way that the bond strength exceeds the cohesive strength of the polymer.

The various standard methods of pretreating polyethylene are, briefly, as follows:

(i) Chemical treatment,[1] using for example chromic acid solution. This method is slow, messy, and tends to produce discoloured, crazed surfaces. It is easy to overtreat, so that although the surfaces produced have very low contact angles with water, high bond strengths are not obtained.

(ii) Flame treatment,[2] in which the polyethylene surface is treated directly with a gas/air flame. Although very rapid, and capable of giving very high bond strengths, this method poses severe problems of control, when using polyethylene film.

(iii) Hot air treatment,[3] which is both slower and less effective, in general, than flame treatment.

(iv) Corona treatment.[4] This is the normal method of treating polyethylene film and is rapid and effective. The polyethylene surface is exposed to the corona discharge in air at atmospheric pressure.

Recently the use of discharges other than corona, operating at reduced pressure, have been reported. These have included the use of atomic oxygen generated at radio-frequencies,[5] glow discharge in air[6] and radio-frequency discharge in inert gases.[7]

These do not seem, so far, to have gained widespread commercial

*Now at Formica International Limited.

application. It is not certain how effective they are compared to the standard methods, but it is likely that they are rejected not because of lack of efficacy, but because of the necessity for a vacuum set-up.

The present work is concerned with the use of glow discharge, which can be defined for this purpose as a stable, luminous discharge, operating at pressures of about 0·1–20 torr, at current densities of up to 20 mA/cm^2. The discharge can be operated by alternating or direct current, but this work has been exclusively concerned with frequencies of 1 MHz, because of the impedance of the polyethylene films. At this frequency, voltages of a few hundred volts are used. The glow discharge is initiated by a spark type breakdown which provides free ions, which bombard the electrodes, causing the liberation of secondary electrons, which cause further ionization. Under the right conditions the system rapidly stabilises to give the typical glow, which often contains coloured lamellae parallel to the electrodes.

Experimental

Apparatus for discharge treatment

The apparatus used, shown schematically in Fig. 97, consisted of a vacuum chamber which could be evacuated by means of a rotary oil

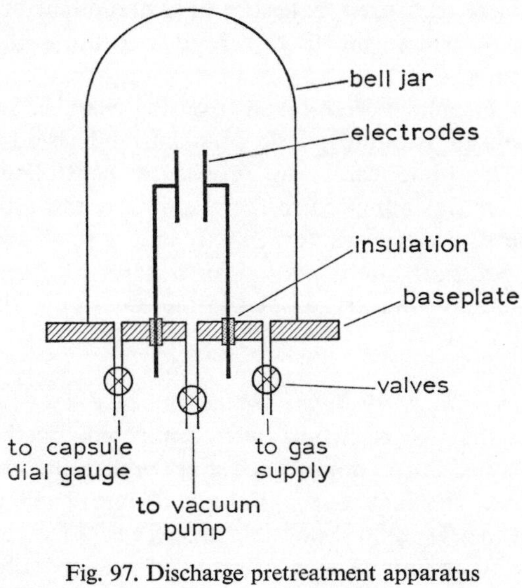

Fig. 97. Discharge pretreatment apparatus

pump (down to 0·1 torr) and a silicone oil diffusion pump (down to 10^{-7} torr). The chamber was fitted with suitable pressure measuring devices (Pirani-Penning gauges for lower pressures, and a capsule dial gauge for pressures up to 5 torr). Gas could be admitted to the desired pressure through a needle valve.

The generator used supplied power at 1 MHz, via electrically insulated leads through the base-plate, to the electrodes, which consisted of massive steel blocks.

Technique of discharge treatment

The samples consisted of a Phillips type high density polyethylene, 4 mm thick, measuring 8 cm × 7 cm. They were placed over the facing surfaces of the electrodes, and held in position by adhesive tape round the edges. The sides and back of the electrodes were heavily covered with insulating material, as were the electrical leads, in order to prevent stray discharges. The electrodes were placed at the required distance apart on an insulating support, and the chamber then pumped down to less than 10^{-4} torr. After admitting the pre-treatment gas to the required pressure the discharge was struck, and run for the required time using a time switch. Discharge currents and voltages were monitored during the run by means of valve voltmeters. When using the very short treatment times, the discharge current and voltage were measured at leisure in a preliminary experiment, and the actual experiment then carried out using the identical generator settings.

Normally the samples were taken from the chamber immediately after admitting air to atmospheric pressure, and tested or bonded immediately. In some cases, to provide a controlled oxidizing environment, air was admitted to the chamber to atmospheric pressure after the discharge was turned off, and a small ozone lamp, rated output 5 mg per hour, was run for a standard time, inside the bell-jar, before removing the samples in the usual way.

Sample testing

(a) *Bond strength.* After pretreatment, the two identically treated samples from the two electrodes were laminated together using a catalysed epoxy adhesive, applied at room temperature and cured at 40°C overnight. The laminate was then cut into 2 cm wide strips and subjected to a T-peel test on an Instron Tensile Tester at a cross-head speed of 50 cm/min. The results quoted are generally the mean of three. The adhesive formulation was altered from time to time, so that it is not always possible to compare the results of different series of experiments, except in general principles. Within any one series, however, every effort was made to ensure consistency.

(b) *Contact angle*. The contact angle of a sessile water drop of known volume on the treated surface was measured by a microscope fitted with a goniometer eye-piece. The advancing contact angle was taken after allowing fifteen minutes to elapse from the application of the drop.

Results and discussion

Effect of gas pressure

Trial experiments showed no systematic effect of oxygen pressure in the range 1–3 torr. A pressure of 3 torr was selected for the majority of the subsequent experiments.

Electrode separation

The results are shown in Fig. 98. The conditions were as follows: discharge current 100 mA, voltage 350 volts, discharge time 30 sec,

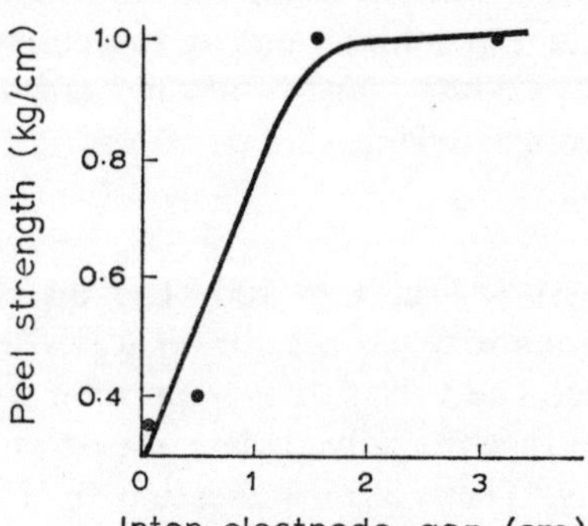

Fig. 98. Effect of electrode spacing

oxygen pressure 3 torr. It can be seen that the peel strength obtained increases with increasing electrode separation up to about 1 cm and thereafter remains constant. A separation of 2 cm was chosen for all subsequent experiments. The reason why the treatment is not so effective at small separations is not clear, but it is probably associated with the observation that at these low separations the discharge itself was very irregular and erratic, being prone to sparking.

Discharge current

The results are shown in Fig. 99. The discharge voltage varied between 330–410 volts, depending on the current, the discharge time

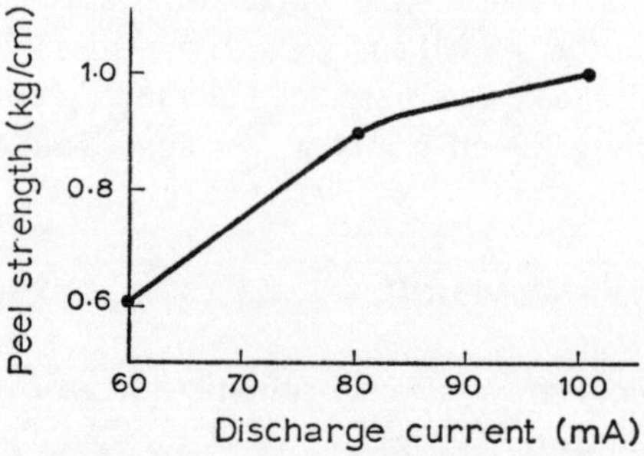

Fig. 99. Effect of discharge current

was 30 sec and the oxygen pressure 3 torr. Bond strength increases with discharge current, but above 100 mA seems to be levelling off. There is a limitation in the maximum discharge current that can be used; if too high the heating effect becomes marked, so that the polyethylene becomes wrinkled, and at very high currents (> 10–15 mA/cm^2) serious arcing and sparking can occur, leading to puncturing of the polyethylene.

Discharge time

The results are shown in Fig. 100. The discharge current was 100 mA, at 350 volts, oxygen pressure 3 torr. Bond strength rises with exposure time. The values above approximately 1·6–2·0 kg/cm width simply mean that the polyethylene was stretching and tearing, rather than peeling. There is an indication of a levelling off after about 1 minute treatment, followed by a further rise which leads to bond strengths exceeding the cohesive strength of the polyethylene.

The effects of time and current upon bond strength are connected, in that higher current densities produce a given bond strength in shorter times.

Although it is difficult to be sure, because of the different methods of test, different materials, etc., used by different authors, it seems that the type of treatment described here is rather faster and more effective than others described in the literature, with the possible exception of the work of Schonhorn and Hansen.[7] For example,

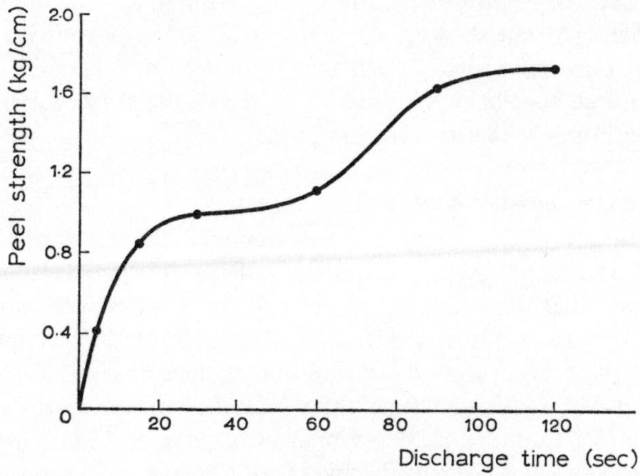

Fig. 100. Effect of discharge time

Hansen and co-workers[5] passed oxygen through a radio-frequency coil at 1 torr pressure, generating atomic oxygen, which was allowed to impinge upon various polymers. No times to produce bondable surfaces were quoted, but the time axes on their graphs often went up to 90 minutes.

Elsewhere[6] a glow discharge device for pretreating polyethylene bottles for printing or lacquering, as an alternative to more conventional methods, has been described. A diffuse glow at 200–400 V (a.c.), and 0·5 torr air pressure, was allowed to impinge upon the polyethylene. Treatment times of 10–15 sec were quoted, and it should be noted that the pretreatment level required for adhesive bonding is generally much higher than for printing.

The reason why the technique described here is quicker and more effective than others is felt to lie in the relative configuration of polyethylene film and electrodes. In this work the polyethylene completely screens the electrode, so that it is the polyethylene surface which receives the direct ion bombardment and emits the secondary electrons that support the glow. In other work, the plasma has been generated elsewhere and simply allowed to impinge upon the polymer.

To test this hypothesis, some experiments were carried out using our own apparatus, in which the polyethylene, rather than being in contact with the electrodes, was simply suspended between them.

Under these conditions negligible bond strengths were developed. Also some experiments were carried out in which polyethylene was exposed to an electrodeless discharge in oxygen at 1 torr for 30 sec. The discharge current was 11 mA and the voltage 1 400 volts. Again negligible bond strengths were produced.

Contact angle measurements

Unfortunately, contact angle measurements are not available for most of the experiments described earlier. However, there are indications that the method described here does not normally produce very low contact angles with water. Rather, a contact angle low enough to ensure good wetting-out of the adhesive is obtained quickly and the contact angle then remains constant with further exposure. For example, an experiment was carried out in which film was exposed to the oxygen discharge at 150 mA, 450 volts, 1 torr oxygen pressure for varying times. The results are shown in Table 29.

TABLE 29

Discharge time (sec)	Contact angle (degree)	Peel strength (kg/cm)
20	35	0·8
40	36	1·4
60	35	1·4

It is suggested that the reason for the attaining of a limiting value of contact angle is the bombardment of the polyethylene surface. Possibly the free radicals formed on the surface react with oxygen to form, for example, surface carbonyl groups, which on subsequent bombardment are broken up and eliminated as volatile oxides of carbon. A point may be reached at which the rate of formation of surface carbonyls equals the rate of loss.

In the work described earlier on allowing plasma to impinge upon the polyethylene, much lower contact angles were obtained, despite the low bond strengths. In the case of the electrodeless discharge, angles as low as 24° were observed. Also, on carrying out discharge in the inert gas argon, and allowing subsequent oxidation in the atmosphere, contact angles obtained were so low as to be unmeasurable.

From the work reported here, it appears that a low contact angle is not by itself sufficient to ensure good bonding. Clearly the physical state of the polymer surface is important too.

The physical state of the surface

It seems likely that in glow discharge pretreatment, the bombardment initially leads to the formation on the surface of active sites, such as free radicals, and ions. Presumably these can then further react by chain scission, or with oxygen to form peroxy radicals and then peroxide groups or carbonyl groups, or by loss of hydrogen to form unsaturated groups, or by cross-linking with neighbouring chains.

The kink in the discharge time plot (Fig. 100), could be the point at which the heating effect of the glow starts to play a part. Up to this point the wettability, and possibly the cross-linking of the amorphous part of the polyethylene, are increasing to a maximum. As these effects start to level out, the heating effect of the glow upon the polymer surface becomes significant, increasing the rate constant of cross-linking and allowing the polymer chains, especially in the crystalline areas, to become sufficiently mobile to diffuse together and cross-link. Also the branching and unsaturation produced will inhibit crystallization on recooling,[8] leading to the formation of a cross-linked gel, of low crystallinity, which will be strong, easily penetrated by diffusing adhesive molecules, and readily wetted by adhesive solutions.

One of the drawbacks of this apparently ideal situation is the unattractively long times required to produce the effect. Also, the ablative effect of prolonged discharge may result in a type of stress cracking, [5, 9] and the heating may damage delicate substrates.

Therefore it was decided to attempt to improve the bonding at short discharge times.

Short duration treatments

A series of experiments was carried out using the following conditions: discharge current 150 mA, voltage 360 volts, oxygen pressure 1 torr. After completion of the discharge, air was admitted to the chamber to atmospheric pressure, and the ozone lamp run for 1 min. The object of the ozone was (i) to obtain more uniform and reproducible oxidation conditions, and (ii) to attempt to improve the bond strengths obtainable.

The bond strengths obtainable by the use of ozone were slightly improved, as shown by Fig. 101, which illustrates the results of some subsidiary experiments to establish the effect of ozone. The ozone treatment rapidly attains a limiting value, of about 0·7 kg/cm, compared to that obtained with no ozone treatment of about 0·5 kg/cm.

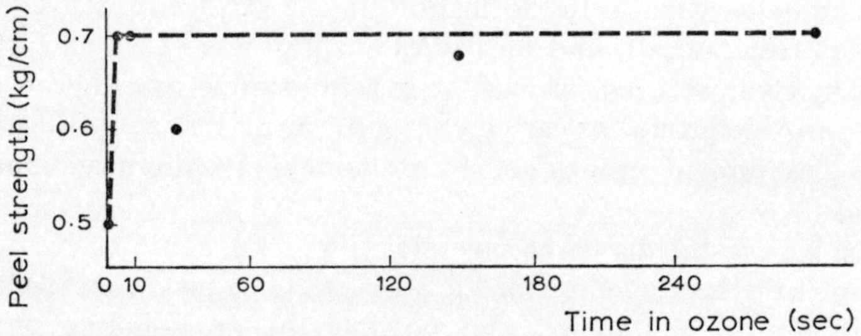

Fig. 101. Effect of ozone exposure

The effect of reducing the time of discharge to the lowest time obtainable experimentally with our set-up is shown in Fig. 102. As can be seen, at very low times the peel strength is in fact increased. It is suggested that the explanation is that the initial effect of bombardment is to cause chain scission, leading to a reduction in the cohesive strength of the polymer at the interface. Later the cross-

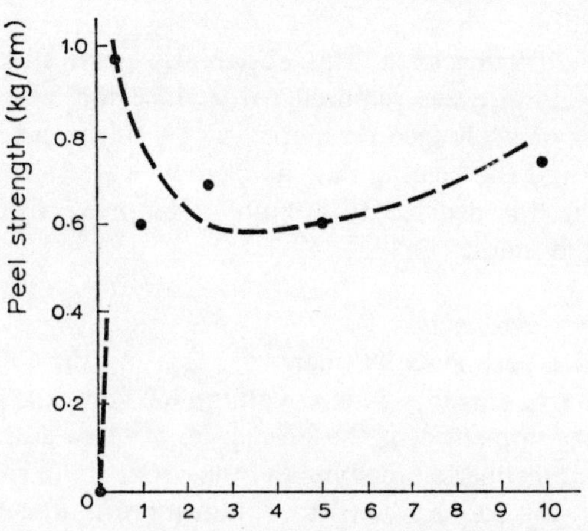

Fig. 102. Effect of discharge duration at short times

linking reaction takes over as the surface starts to heat up, and the peel strength again increases as the interfacial material is strengthened. The bond strengths obtainable after very short times are probably not so high as after prolonged discharge, but are nevertheless very satisfactory.

This effect will only be observed for a polymer which has a strong surface initially. The method described here of producing short discharge times is probably not the best, since a proportion of the apparent time will be taken in initiating and stabilising of the glow. The use of moving webs of polyethylene would be better.

Discharge in inert gases

As mentioned briefly earlier, some experiments were carried out using non-oxidizing gases to support the discharge.

These were hydrogen, at a discharge current of 150 mA and 360 volts at 1 torr, and argon under identical conditions. In both cases, discharge was followed by ozone exposure for 1 minute. The results are shown in Table 30.

TABLE 30

Gas	Discharge time (sec)	Peel strength (kg/cm)
Argon	10	0·60
	5	0·55
Hydrogen	10	0·85
	5	0·70
	1	0·90

Again the tendency seems to be for slight rise, in the case of hydrogen, in peel strength at the lowest times. It should be noticed that hydrogen is considerably more effective than oxygen as the discharge gas. Argon is slightly better.

Conclusions

It is difficult to be sure how the bond strengths quoted here compare with those of other workers, as mentioned before. One of the reasons is that the morphology of the polythene surface, which can be profoundly affected by the method of preparation of the sample, is very important in this sort of work. However, this technique does produce bonds exceeding the cohesive strength of the polymer at exposure times, depending on the discharge parameters, which can be as short as half a minute. At extremely short discharge times, less than 1 sec, high bond strengths are again obtained. The upper limit of the bond strength obtainable in this way has not been established.

The general conclusion reached here is that for the production of polyethylene surfaces capable of giving very strong adhesive bonds, the surface should be polar enough to be adequately wetted by the adhesive solution, and strong enough not to fail cohesively in a weak surface layer. It is not necessary to go to very low contact angles in order to get good bonds, and in fact in many cases too low a contact angle can lead to poor bond strengths because it is obtained at the expense of the strength of the interface. Also, too low a contact angle on the treated surface can render the interface moisture sensitive.[10]

Schonhorn and Hansen[7] have reported the production of good bondable polyethylene surfaces by the use of radio-frequency activated species in inert gases, without any change in the wettability of the surface. Using our methods, we have never observed this, perhaps because as mentioned earlier, although the treated material, after discharge in, say, hydrogen, has a contact angle of approximately 90° immediately on removal from the chamber, this rapidly falls to immeasurably low values during the course of 10–20 min exposure to the laboratory atmosphere. This is in general accord with the observations made by Bamford and co-workers[11] on the life-times of free radicals produced on the surface of polymers by a Tesla coil discharge in hydrogen. If good bonding is produced without change in wettability, it is possible that these chemically very active surfaces are actually grafting with the adhesive.

References

1. Horton, P. V. U.S. Patent. **2,** 558, 134, 1954.
2. Kritchever, M. F. U.S. Patent. **2,** 648, 097, 1953.
3. Kreidl, W. H. U.S. Patent. **2,** 632, 921, 1953 and **2,** 746, 084, 1956.
4. Von der Heide, J. C., and Wilson, H. L., *Mod. Plast.*, **38,** 199, 1961.
5. Hansen, R. H., Pascale, J. V., de Benedictis, T. and Rentzepis, P. M., *J. Pol. Sci.*, **A3,** 2205, 1965.
6. Anon. *Die Neue Verpackung.*, 1366, October 1965.
7. Schonhorn, H. and Hansen, R. H., *J. Appl. Pol. Sci.*, **11,** (8), 2062, 1967.
8. See for example, Chapiro, A., in *Radiation Chemistry of Polymeric Systems*, p. 396. Interscience, New York, 1962.

9. Mantell, R. M. and Ormand, W. L., *I. and E. C. Prod. Res. and Dev.*, **3**, (4), 300, 1964.
10. Bright, K. and Malpass, B. W., Paper presented at the British Surface Physics Conference, Churchill College Cambridge, January 1967; and Bright, K. and Malpass, B. W., *European Polymer Journal*. In Press.
11. Bamford, C. H., Jenkins, A. D. and Ward, J. C., *Nature*, **186**, 712, 1960.

Evaluating the Performance Qualities of Gummed Paper Tape

T. L. WRIGHT, A.C.T.(Birm.), A.P.I.
THE GELATINE AND GLUE RESEARCH ASSOCIATION

Introductory

The object of this paper is to discuss the variables which affect the performance of gummed paper tape. Particular reference will be made to a recently developed testing instrument designed to examine the adhesive properties of gummed tape. Its capabilities will be illustrated by describing some of the investigations carried out at both the Gelatine and Glue Research Association and at Darling and Company, Chicago, into factors influencing the rate of strength development of gummed paper tape.

The gummed tape as commercially available is an adhesive coated paper which is usually supplied in rolls of various dimensions usually in the order of 2–6 in wide by 600 ft long. A variety of water-soluble adhesives may be used for this application although the work which will be described has been based on animal glue. The tape is used for the packaging of cartons, boxes, parcels, etc. The colour and surface decoration can be suited to the customer's requirements and can be designed to provide a useful source of advertisement.

The gummed tape industry has always provided a substantial outlet for animal glue and continues to do so despite competition from cheaper sealing media. Although a breakdown on applications for animal glue in this country is very difficult to obtain, some recent figures[1] for the United States have been published and these are shown in Table 31. It will be seen from this that well over a third of the animal glue produced goes into the manufacture of gummed tape. A figure in excess of 60 000 tons of glue would be an approximation as to the annual world consumption for this application.

In view of this large market for animal glue the Research Association is keenly interested in ensuring that the high quality of animal glue tapes is maintained. In order to accomplish this objective an investigation into the factors affecting the performance of gummed paper was instigated. It was decided that performance could be best

TABLE 31

Sales to Consumer Industries of Natural Glue and Technical Gelatine

	%
Gummed tape and gummed paper	38·3
Paper manufacturers	9·1
Abrasives	8·9
Sizing	6·0
Furniture and woodworking	5·2
Adhesives and glue manufacture	3·0
Cork and cork products	2·7
Matches	2·4
Printers rollers	1·5
Others	21·7

judged by speed of tack or strength development of the wetted adhesive.

Methods by which tack development can be measured and the influence that various factors have on rate of tack development will be discussed later, but firstly the preparation of gummed paper tape will be briefly described.

Raw materials

Paper

The papers used in the manufacture of gummed tape are principally of the kraft variety and they are usually characterised by the weight per unit of surface area. The range covers light tissues of 40–50 g/m^2 up to the heavy reinforced laminations of 250–300 g/m^2.

Various quality control tests[2] such as weight, thickness, smoothness, tearing and bursting strength are carried out on the rolls of kraft paper prior to their despatch to the tape manufacturer. Assuming these properties are satisfactory it is necessary to ensure that the paper is free from formaldehyde[3] which might have been incorporated into the paper to improve its wet strength. Formaldehyde, even in very small amounts, will cause cross-linking of the glue and consequently a loss of tack of the tapes.

Adhesives

As paper is characterised by its weight per unit area, animal glue is usually described by the strength of the gel formed at a standard glue concentration. The strength, in grams Bloom, is recorded as the load required for a cylindrical indentor to depress the gel by

4 mm.[4] For the preparation of gummed paper tape, animal glue of 60 to over 200 g Bloom may be used. It has been suggested[3] that for sealing tape, the optimum strength of the gel, when performance and cost are considered, is around 110 g. This would appear to be somewhat higher than that used in British Industry[5] as the demand for natural glue produced in this country has centered on a gel strength around 80–100 g Bloom. Conversely, European tape manufacturers use natural glue of 130–150 g Bloom.

Manufacture

Gummed paper tape is manufactured by coating an adhesive solution on to a paper and drying either over a hot drum or by means of a stream of hot air. Steps are taken to control the concentration of the adhesive solution, the adhesive loading and the moisture content of the finished tape. All of these factors influence the quality of the final product.

Concentration of a hot water solution of the glue is maintained at the highest level possible without causing streak-marks to ensure maximum tacking rate. Concentrations of 45 to over 60 per cent glue solids, depending upon the characteristics of the gumming head, are employed, whilst temperatures range from 60°–75°C.

The glue loading is controlled at an optimum level to ensure a uniform coverage and satisfactory thickness on the surface. Depending on paper weight, 20–35 g/m^2 are applied commercially.

Moisture is held at an optimum level, which is in equilibrium with an atmosphere of about 20 per cent relative humidity to ensure most satisfactory performance and shelf life of the tape. Such moisture contents range around 3–5 per cent. It should be emphasised that quoted moisture contents refer to the whole tape and not just the glue component.

The finished tape is generally passed over a breaker bar to crack the coating in order to correct natural curl.

Straight glue is used to obtain coatings exhibiting highest rate of tacking. Where desirable, glue is extended with dextrin to reduce the cost of the coating. It produces tapes having a slower tacking rate, but giving satisfactory performance quality for the uses intended.

Rapid drying of the adhesive solution onto the paper ensures that the glue passes directly from a sol to a dried film without at any time becoming a gel. It has been shown by X-ray diffraction studies[6] that hot (> 60°C) dried glue films which are rapidly chilled and stored at low humidities are amorphous and will readily absorb moisture. However, when these films are exposed to damp atmospheres the glue changes to the crystalline structure similar to that exhibited by cold dried glue films. The crystalline form has a lower swell ratio and, therefore, will absorb moisture at a slower rate, consequently the glue takes longer to dissolve and strength development is retarded.

Testing of gummed tape

The performance quality of a gummed paper tape depends on its ability to adhere after moistening and applying to the substrate coupled with the permanency of the bond when dry. The measurement of these properties is somewhat complex in view of the number of variables involved. The performance will depend on (1) the amount of water used for moistening, (2) the porosity of the substrate and (3) the open time, i.e. the time between moistening and applying the tape.

There are a wide variety of testing instruments available for determining the wet tacking rate of gummed tape. They can be divided broadly into three categories according to their method of test, namely shear, adhesion and peel.

The first two test methods as typified respectively by the McLaurin and Neubronner[7] testing instruments both suffer the same drawbacks. Firstly many tests are necessary to obtain a profile of strength development and secondly it has been shown[8] that wide variations occur in the water loading during testing. This latter factor will naturally lead to reduced reliability of test results.

The final mode of test is peeling and there appears to be two main contenders for this field. The first is the Fipago tester[9] made in Germany by the Ströhlein Company. Basically the principal of the operations consists of moistening a test strip and pressing it against a substrate. One end of the test strip is then pulled from the base under gravity and the maximum resistance to peel is recorded.

Again it has been shown[8] that the water-loading is variable and also many tests have to be performed to characterise strength development.

The Research Association considers that these problems have been overcome in the Werle Tack Tester[10] named after its designer, Mr E. E. Werle of the Darling Co., Chicago, and manufactured by the Thwing Albert Instrument Company, Philadelphia, U.S.A.

The main feature of this instrument is that it controls all the factors involved in testing and automatically provides a profile of the rate of development of cohesive strength in the film of adhesive between two substrates. As applied to gummed paper the moistened coating is the adhesive. An outline of the Tack Tester is shown in Fig. 103.

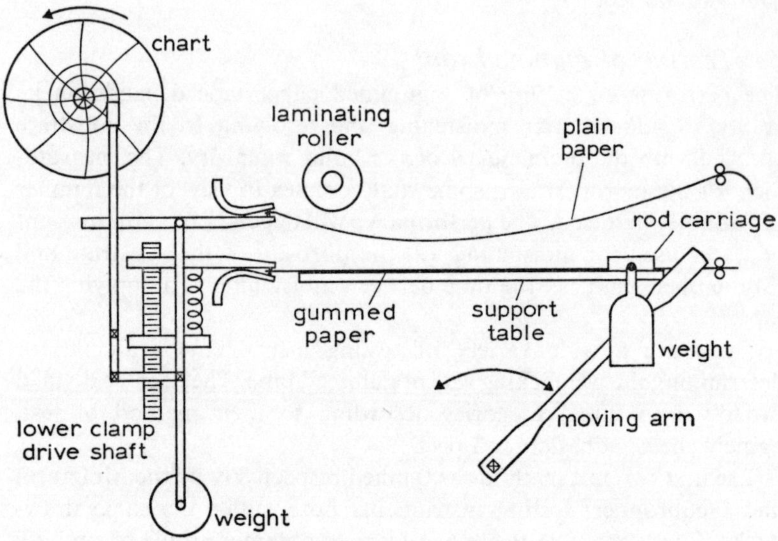

Fig. 103. Werle tack tester

The gummed tape under test is placed on the table and held in position by the lower jaws. A rod, which will be described in more detail later, is placed in the carriage and held in position by weights. This ensures that the rod is maintained in contact with the gummed surface.

A strip of plain paper is clamped in position by the top jaws and a roller is positioned by solenoid clamps. The open and closed times are set and a chart, which is described more fully later, is placed in position. Approximately $1\frac{1}{2}$ ml of water is placed on the gummed tape in front of the rod carriage and the 'start' button is pressed. A pair of arms move the rod carriage across the lower tape, thereby spreading the water. At the end of their movement, which takes 0·7 sec, recording is commenced on the chart. After the pre-set open time has elapsed, the roller drops and the arms move it across the tapes in the reverse direction from spreading, thereby causing lamination of the two tapes. At the end of its movement the roller causes the release of the rear end of each tape.

At the end of the pre-set closed time, the table supporting the tapes drops and causes the clamp holding the gummed tape to fall to take up the slack. Immediately the bottom clamp is driven down a shaft at a constant speed of 1 in/sec, thereby causing peeling of the two tapes. As the peel resistance increases the force is transferred to the top jaw which moves a shaft connected to a pendulum to which is attached a pen. Movement of this pen records resistance to peel on the chart thereby providing a complete profile of strength development.

There are certain features of this instrument which need to be emphasised. Firstly, following the initial application of the water, the operation of the machine is completely automatic so that variation in test results obtained by different operatives is reduced to a minimum.

Secondly, the manner of wetting the tape is such that an even water thickness is applied to the entire test area. The spreader rods which disperse the water are made from $\frac{1}{4}$ in smooth steel bars around which wire is wound. A range of rods is available of different wire diameter to control water loadings between 7–27 g/m^2.

The charts used to record peel resistance are graduated in grams and seconds. Before each test the chart is set back from zero by the open time to allow direct reading of the resistance to peel. The figures recorded are the time (to the nearest 0·1 sec) to reach 20 and 100 grams peel resistance. The 20 g result represents the initial grab characteristics of the tape which is sufficient to hold it in place. The 100 g result represents the strongest cohesive bond attainable before fibre tear commences.

Irregular lines occurring above the 100 g level are attributed to areas of varying density common to most kraft paper.

The open time, i.e. the time between wetting of the adhesive and lamination, is variable from 0·7 sec to 2 min in tenths of seconds and the closed time can be similarly varied between 1·2 sec and 2 min.

In addition to testing the rate of tack development, other properties of gummed tape can also be examined with this machine. For example the open time quality or the ability of the tape to adhere after long open times can be determined.[12] This is performed by varying the open time in stages between 2 and 30 sec and removing the laminated tapes from the machine before the closed time has elapsed and peeling occurs. The laminated samples are then allowed to dry under standardised conditions before separating the tapes with a razor blade. The area of fibre tear is expressed as a percentage of the exposed area.

An assessment of the degree of curl exhibited by a wetted tape can be made by wetting the gummed tape then, using no top paper, cutting the tape from its clamps so that it rests freely on the support plate. A guide to the curl of the tape can be ascertained by measuring the height of arc produced after a standard time. This property is of particular importance in industry when long open times are involved.

Factors controlling rate of tack development

Darling and Co. experimental work

Extensive research has been carried out into the factors which govern strength development of gummed paper tapes and the work is summarised in an article by Werle.[3] Tests carried out with the McLaurin Tester revealed the following effects of manufacturing variables on the performance quality of animal glue tapes.

(a) *Adhesive concentrations*

Fastest tacking rate is obtained by coating with the highest concentration of glue solution as may be seen by referring to Fig. 104.

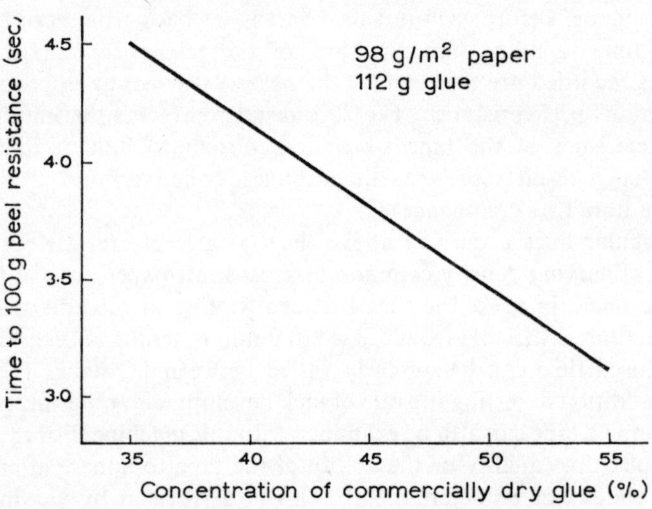

Fig. 104. Effect of glue concentration on rate of tack development

(b) *Optimum moisture content of tape*

Tapes, when made, having a moisture content in equilibrium with an atmosphere of 20 per cent relative humidity, exhibit excellent shelf life and very satisfactory performance qualities. For a tape having 25 g/m^2 of animal glue coated onto 98 g/m^2 paper, such moisture content comes close to 4·0 per cent.

Tapes having considerably higher moisture contents, say in the region of 7 per cent, are liable to lose their tacking quality over relatively short periods of storage, especially at low temperatures. Tapes should be stored on high shelves (warmest room area) for best shelf life.

(c) *Effect of Bloom strength of glue on tacking rate*

Tacking rate is accelerated with increasing gel strength.

(d) *Effect of extending animal glue with gumming dextrin*

Tacking rate is retarded with increasing additions of dextrin.

(e) *Effect of glue loading*

The actual weight of coating on the tape is less important than the amount which remains on the paper surface and the uniformity with which it is applied.

All of these factors are important in the manufacture of gummed tape. It is the concern of the manufacturer to take advantage of this knowledge to produce top quality tapes from the materials available.

To examine the relationship between Werle's results with the McLaurin tester and information attainable with the more sophisticated Werle Tack Tester, an extensive programme of work based on a similar approach to that adopted by Werle was carried out at the Research Association.

G.G.R.A. experimental work

Tapes were prepared in our laboratories for investigating some of the variables listed above using the Werle Tester.

Besides spreading water on to gummed surfaces the machine can be used to prepare gummed tape by spreading the adhesive on to plain kraft paper. The animal glue is applied to the paper and is spread with a pre-heated rod. The wet tape is taken from the machine and dried in an air circulating oven for $\frac{1}{2}$ min at 105°C. By this procedure a tape with an even film thickness of adhesive is obtained. The disadvantage is that only small lengths of gummed tape can be prepared so that a tape making machine has been designed at the Research Association. An outline is shown in Fig. 105. Kraft paper

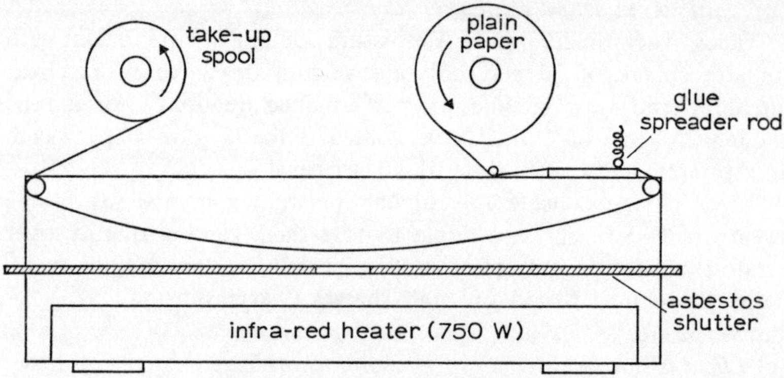

Fig. 105. Laboratory tape coater

is drawn from the roll across a coating table where hot animal glue is applied. The glue is coated on to the tape by means of one of the rods from the tester hence ensuring a known even glue loading. The wet tape then passes through a chamber where it is dried by means of an infrared heater. The tape is finally wound off on a spool which is attached to a constant speed motor. In this manner sufficient lengths of tape were prepared for adequate testing from any adhesive formulation made under various preparative conditions.

Effect of gel strength on tacking rate

Perhaps the first consideration by a tape manufacturer is what adhesive he should use. This decision will be to a large extent governed by the application for which the tape is designed. For certain applications where little strength is required, e.g. postage stamps, then the cheap starch-based adhesive can be satisfactorily employed. When, however, greater strength is required, for example in packaging and box sealing, then the case for using animal glue is far stronger.

The problem then presents itself as to which animal glue to select from the wide range of gel strengths available. As a first step in an investigation of the effect of glue properties on tack development a series of tapes were prepared by coating a constant weight of various gel strength glues on to 98 g/m^2 kraft paper. This paper weight is typical for packaging applications as is the open time of 2 seconds and water and glue loading of 17 and 27 g/m^2 respectively that were used throughout this investigation. A graph of the results is shown in Fig. 106. From this graph it can be seen quite clearly that as gel strength rises, i.e. with improvement of glue quality, then the rate of tack development is considerably accelerated.

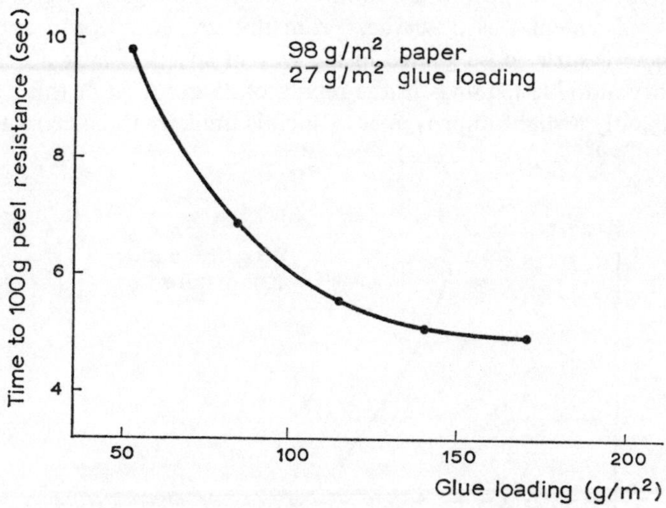

Fig. 106. Effect of glue quality on rate of tack development

A bone glue with a gel strength value of 102 g Bloom at 12½ per cent concentration was adopted as the standard. The main reasons for this was that this strength is typical for that used by British Tape Manufacturers.

The next stage of the investigations was to establish a constant glue concentration in order to prepare a standard tape with which to study other variables. As indicated in Fig. 104, as glue concentration is increased, the tacking rate is faster. There is obviously a maximum glue concentration above which it is impossible to produce a satisfactory coating. The limit with the standard glue using the tape spreader was found by trial and error to be a solution containing 48 per cent commercially dry glue (i.e. 40 per cent solids) coated at 60°C. In commercial operation, temperatures and concentrations can be increased to a degree consistent with the characteristics of the gumming head.

Effect of glue loading on tacking rate

The next feature of tape manufacture to be examined was quantity of glue applied to the tape. Using the same glue as in the previous work, varying quantities were applied to a standard kraft paper. The rate of tack development with increasing glue loading is shown in Fig. 107. At very low glue quantities, i.e. in the region 10 g/m², no tack development was observed. From this graph it would appear that there is little to be gained in the way of speed of tack development beyond glue loadings in the region of 25 g/m². After this point there is only a slight improvement which is unlikely to be economically justified.

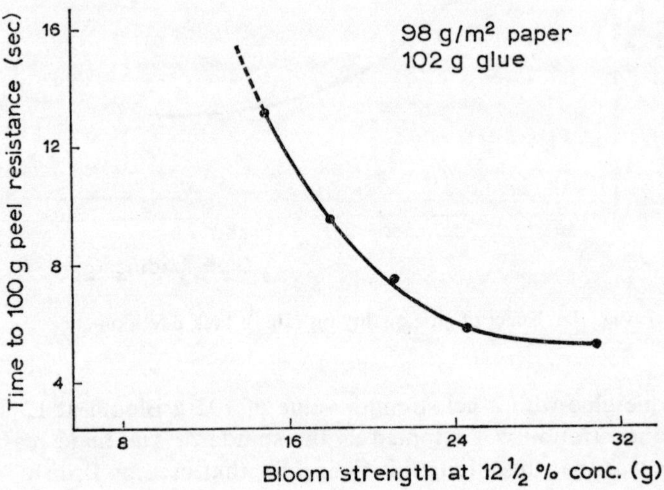

Fig. 107. Effect of glue loading on rate of tack development

It should be understood that the performance of a tape depends not so much on the amount of loading of coating as it does upon the amount that remains on the surface. Therefore, if one applies highest concentration of solution, there is less penetration and more remains on the surface. Consequently, relatively lower loadings can be tolerated if penetration is held to a minimum.

Effect of water loading of tacking rate

Having discussed the effect of glue in both its quality and quantity the next and most important factor is the influence of moisture.

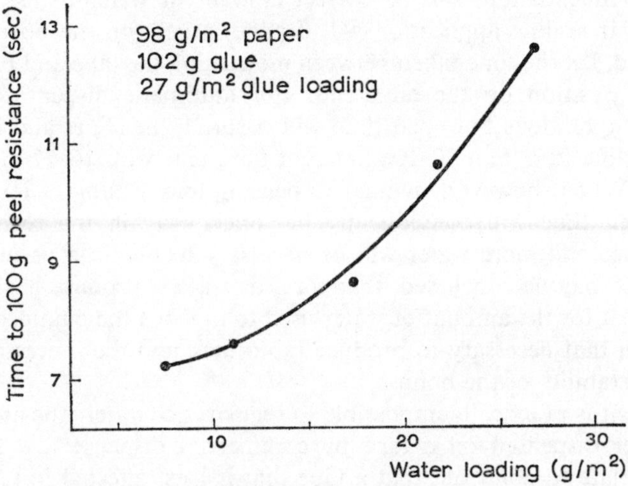

Fig. 108. Effect of water loading on rate of tack development

Both Werle[13] and Niewind[14] have demonstrated that using different testing instruments, the quantity of water used to moisten a gummed paper greatly influences the rate of tack development and permanency of seal. If one first considers simply the change in rate of tack development with varying water applications then results of the type shown in Fig. 108 will be obtained. It can be seen quite clearly that the fastest tacking rates are obtained at the least amount of moistening. This will be due to the more rapid bond being formed in the cases where there is less water to be absorbed by the substrate. Judged on this information alone it would appear that the best results are to be obtained by using the minimum amount of water. The drawback, however, is that when these bonds are examined after standing overnight, those made with too little water have either already fallen apart or do so at the slightest stress. This was attributed to insufficient penetration of the adhesive into the adherend. At higher water loadings satisfactory permanency of bond is obtained. This may be demonstrated with the Werle Tack Tester by preparing laminations with the various spreader rods and before peeling takes place, the laminants are removed from the instrument and allowed to dry overnight. The top tape is then removed by scraping with a razor blade and the area of fibre tear in relation to the total area examined is expressed as a percentage.

Recommendations on the correct amount of water to use for a particular sealing application will depend largely on the open time involved, i.e. the time taken between moistening the tape and placing it into position on the adherend. For automatic dispensing and sealing operations the open time will naturally be short and it will be possible to obtain 90–100 per cent fibre tear with 16–17 g/m^2 of water. When, however, manual dispensing and sealing is involved the open time will consequently be longer, say in the region of 20–30 sec and more water will be necessary to obtain a permanent bond. It may be concluded, therefore, that a compromise has to be arrived at for the amount of water used to moisten the gummed tape between that necessary to produce rapid tack and that necessary to ensure stability of the bond.

Since it is practically impossible to regulate accurately the amount of water dispensed on a tape by commercial dispensers, it seems appropriate to point out that a tape that is least affected in tacking speed by variations in such amounts dispensed will perform the best. In this regard, animal glue tapes have been found to be much more reliable than modified starch tapes.

Comparison of adhesives for gummed paper tape

As mentioned earlier animal glue is being challenged as an adhesive for gummed paper tape by various modified starch materials. The major attraction of the modified starch tapes is undoubtedly price. However, there are many other considerations to be made before automatically choosing the cheapest material and it is these factors which need careful evaluation. Where adhesion without loading is required, as with labels for example, then the tack of modified starch adhesives is adequate. On the other hand where there is a definite force to overcome as with a springy cardboard carton then the strength of the animal glue is very much more suitable.

Two criticisms of animal glue have clouded this issue. Firstly there is always the association of ideas that this glue has an odour and in the past manufacturers did not pay sufficient attention to this objection. Today, however, with improved manufacturing techniques, materials with greatly reduced odour levels are available, some virtually odourless.

Secondly the re-wetting quality is argued. As to the wetting quality, uneven distribution of water on a tape's surface is an asset rather than a hindrance. Apparently unwetted areas are indeed wet and contribute fast immediate grab or stick which is characteristic of animal glue. These areas do not affect the completeness of the final bond.

Animal glue packaging tapes have the following main points of advantage over the modified starch tapes:

(1) They curl much less both in the dry state and after moistening.

(2) They show a distinct immediate grab quality upon application compared with the slip exhibited by modified starch.

(3) They are much less sensitive to variations in the amount of water applied. They adhere far more rapidly than modified starch tapes when large amounts of water are applied.

(4) Different productions are more uniform.

The advantages of starch formulations lie mainly in the price per lb and possibly less odour. On differential costing, it should be kept in mind that because of specific gravity differences, 25 g/m^2 of dry glue formulation will give the same film thickness as 31 g/m^2 of starch formulation—thus narrowing the price gap by nearly 25 per cent.

Pressure sensitive tapes are used on a relatively small scale for a few industrial packaging applications, but despite their handling advantages, their price and tendency to deform under tension have limited their progress in this field.

Summary and conclusions

Animal glue for gummed tapes

There is probably no known adhesive that has the fast tacking rate of animal glue. It is possible to tailor-make the tapes to have extremely rapid tacking rate or relatively slow, as required. This can be done merely by selecting the gel strength of the glue and/or by extending the glue formulation with gumming dextrin.

Starch sealing tapes have generally a less versatile level of tacking rate, the quality of which is not always consistent from different lots of starches. This variation in commercial tacking performance is a frequent cause for complaint. Whereas the tacking quality of animal glue is regulated by its gel strength, there is no such property upon which starch can be evaluated.

In the United States several attempts at producing reinforced sealing tapes, that is, paper reinforced with nylon or glass fibre strands, with modified starch coatings were completely unsuccessful. Only animal glue coatings are suitable for such use.

The Werle tack tester

The principles of operation of this machine provides one with the necessary requirements to evaluate the effect of variables involved in any study of a liquid adhesive between two substrates. It can be used to test remoistenable tapes, liquid adhesives on a given substrate or the gluability of a given substrate with different adhesives.

Acknowledgement

The author is most grateful to Mr. E. E. Werle of Darling and Company, Chicago for his editing of this paper and his help throughout this work.

References

1. Courts, A., private communication.
2. Paris, Sesar and Gasteigner, *Adhesion*, **5,** (1), 1; (2), 61, 1961.
3. Werle, E. E., *Paper, Film and Foil Convertor*, pp. 23 and 91, Feb., 1960.
4. British Standard 647: Section 6, 1959.
5. Dunnett, P. S., G.G.R.A. Research Panel Paper 61.
6. G.G.R.A. Abstracts **8,** (1), 1, 1957.
7. Neubronner, C., *Adhesion*, (11), 457, 1964. (*G.G.R.A. Bull.* **16,** (1), 12).
8. Dieth, A. and Schifferr, H., *Adhesion*, (2), 55, 1967. (*G.G.R.A. Bull.* **18,** (2), 18).
9. Strohlein and Co. Tech. Report (Bb505–E).
10. Werle, E. E., Tappi 15th Testing Conf., 1964.
11. Thwing Albert Instrument Co., Tech. Sales Data (229–001), 1964.
12. Draft of Test Procedure for Adhesiveness of Gummed Paper Tape (Tappi Format), Jan., 1967.
13. Werle, E. E. and Driscoll, W., Correspondence, G.I.A., Aug., 1964.
14. Niewind, R., Fipago Cong., May, 1967.
15. Werle, E. E., *Tappi*, **44,** (7), 1961.

Modern Developments in the Study of Interfaces

PROFESSOR R. C. PITKETHLY, B.Sc., Ph.D., F.R.I.C.

B.P. RESEARCH CENTRE, SUNBURY AND THE CITY UNIVERSITY, LONDON

An anonymous article appeared in *Chemical Week* at the end of October last,[1] in which Irving Skeist, editor of the *Handbook of Adhesives,* was quoted as saying, 'Companies don't do basic research in this business. They can't afford to.'

It was argued in this article that, although many industrial technologists in this field believe that the pragmatic approach will continue to succeed as it has demonstrably done in the past, there are areas in which inadequacies and the lack of a coherent theory are apparent. Although, 'there are no signs in the adhesive field of the progress in basic knowledge that has marked such areas as biochemistry, nuclear energy and surface chemistry', I would suggest that the time is ripe for such developments.

A major difficulty in the formulation of useful fundamental research programmes in adhesion is the sheer complexity of behaviour of the practical systems. But this is true also in other fields and it should be instructive to consider the way research has developed in one of these, viz. surface chemistry and catalysis.

Heterogeneous catalysis has often been described as an art and not a science. In spite of the high quality of the experimentation and logic in individual studies, catalysis research as a whole has suffered from lack of control over its materials and this has stemmed from the fact that very little was known for certain about the detailed structure of the surfaces which were studied and consequently of the location or nature of the activity. An indication of this uncertainty may be seen in the many texts which discuss catalytic reactions in the following form without explaining the detailed meaning of X:

$$\begin{array}{ccccccc} CH_2 & - & CH_2 & + & H & \longrightarrow & CH_2 - CH_3 \\ | & & | & & | & & | \\ X & & X & & X & & X \end{array}$$

Powerful techniques have been applied to the problem of bonding of reactants to surfaces, e.g. various kinds of spectroscopy and microscopy, ellipsometry and crystallography; every new invention is pressed into service but in spite of this there is still serious argument over interpretation of apparently simple systems. Consider for example, the continued discussion on the chemisorption and reactions of oxygen and hydrogen on platinum and their use for the estimation of available metal surface.[2–5] However, great advances have been made in recent years in the theoretical and experimental methods and in the understanding of catalytic processes. Not only have these advances led to explanation of the empirical discoveries but have led to and encouraged the discovery of new, interesting and industrially valuable catalytic systems.

The most interesting and significant developments in catalysis are taking place in areas where detailed chemical and physical structure is emphasised, e.g. in catalysis by transition metal complexes,[6,7] the identification of the nature of catalytic sites on oxides and in zeolites,[8–10] in the correlation of the properties of the crystallites of supported metal catalysts with their size,[11–13] exposed edges and faces,[14–17] step structure and concentration of dislocations[18,19] and so on.

The fundamental studies of interfaces, or 'interphases', whether in connection with catalysis, lubrication or adhesion, have already reached the stage where concentrated efforts must be made to elucidate the structure of the surfaces down to atomic detail and the physical and chemical interactions at individual surface atoms. Naturally, integration of co-operative effects will be necessary if large uniform areas or large molecules are involved but the atom's eye view will certainly be necessary if the surfaces are irregular. In the situations where physical interactions rather than chemical bonding are important, much more attention has to be paid to the changes in the structures, atomic spacings, etc., of the solid and liquid at and near the interface and to the properties of surface irregularities.

It is such considerations and particularly our work on the selective poisoning of nickel catalysts by organic sulphur compounds[20] that led us to study single crystal faces as a step towards more precisely defined practical surfaces. Field emission and field ion microscopy[21,22] have thrown much light on the properties of specific crystal faces of refractory metals, and the mobility of atoms and the preferred tracks by which they move over such surfaces. However, for our catalyst studies it was necessary to select techniques which would not be limited to refractory inorganic systems and would permit handling of organic materials at the surfaces. This encouraged us to tackle the construction of Low Energy Electron Diffraction equipment in 1964. LEED has two advantages for this kind of work; first, the slow electrons have low penetration and consequently give information solely from the surface layers, and secondly their wavelength is the same order of magnitude as the inter-atomic spacings in solids. For a general account of this technique, see Ref. 23.

The recent literature contains numerous reports of studies of (i) metals, Pt, W, Mo, Fe, Ni, Cu, Ag, etc.; (ii) semiconductors, Si, Ge, GaAs, etc.; (iii) insulators, Al_2O_3, MgO; (iv) ordered physi- and chemisorbed layers and (v) epitaxially grown thin films. In this paper the LEED technique and equipment will be very briefly described and results will be presented of selected experiments which demonstrate distortions and rearrangements in interphases at the surfaces of single crystals.

Equipment

Our LEED investigations have been carried out in two units. The first apparatus was built entirely at the B.P. Research Centre at Sunbury; the second, a Varian Associates apparatus, was installed there in September 1965. With the support of the Science Research Council a Vacuum Generators' equipment is about to be commissioned at the City University. It combines LEED with quadrupole mass spectrometer, and more sophisticated gas dosing systems are being developed for studying various reactions on metals.

Very low pressures are necessary if clean surfaces are to be obtained and kept uncontaminated. Molecular bombardment on a surface exposed to a gas at 10^{-6} torr for one sec is approximately sufficient to give a monolayer and at 10^{-10} torr, with unity sticking factor, monolayer coverage takes 2–3 hours.

The schematic diagram in Fig. 109 shows a typical LEED apparatus which consists essentially of three sections: pumps, gas inlet system and the LEED chamber. The molecular sieve sorption pumps take the pressure down to 10^{-3} torr and the 140 $l.sec^{-1}$ sputter ion pump provides continuous pumping to 3×10^{-11} torr. High purity gases are stored in glass bulbs attached to the Gas Inlets and admitted, as required, through a bakeable leak valve, b, into the LEED chamber. The crystal is supported in a holder which encloses a tungsten filament heater and this assembly can be moved inside the chamber by means of an external manipulator. An ion bombardment gun,

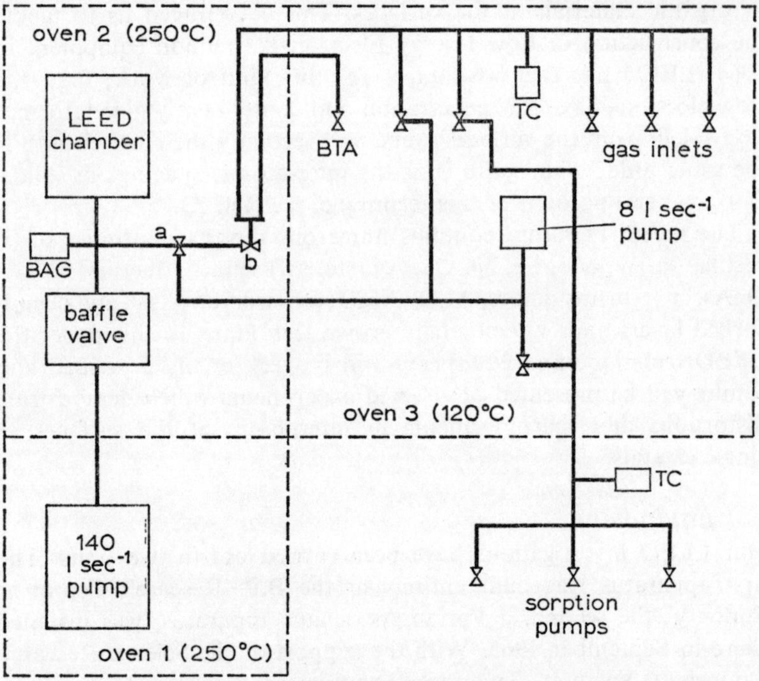

Fig. 109. Schematic of Varian LEED system. a—T valve, b—Leak valve, BTA—Back to air valve, TC—Thermocouple gauge, BAG—Bayard Alpert gauge (*Courtesy Varian Associates Ltd.*)

Bayard-Alpert pressure gauge, baffle valve and the electron optical system are also provided.

A schematic diagram of the electron optics is shown in Fig. 110. A monoenergetic electron beam is directed from the gun on to the crystal. Electrons are back-scattered into a field-free region determined by the walls of the chamber and the first wire gauze grid. The second grid is held at a potential which will reject inelastically scattered electrons but will permit passage of the elastically scattered electrons through to the fluorescent screen which is held at 5 kV potential.

Theory and practice

Simple kinematic theory of diffraction from a plane grating (Fig. 111) and the Bragg relationship (Equation (1)) using λ from Equation (2)

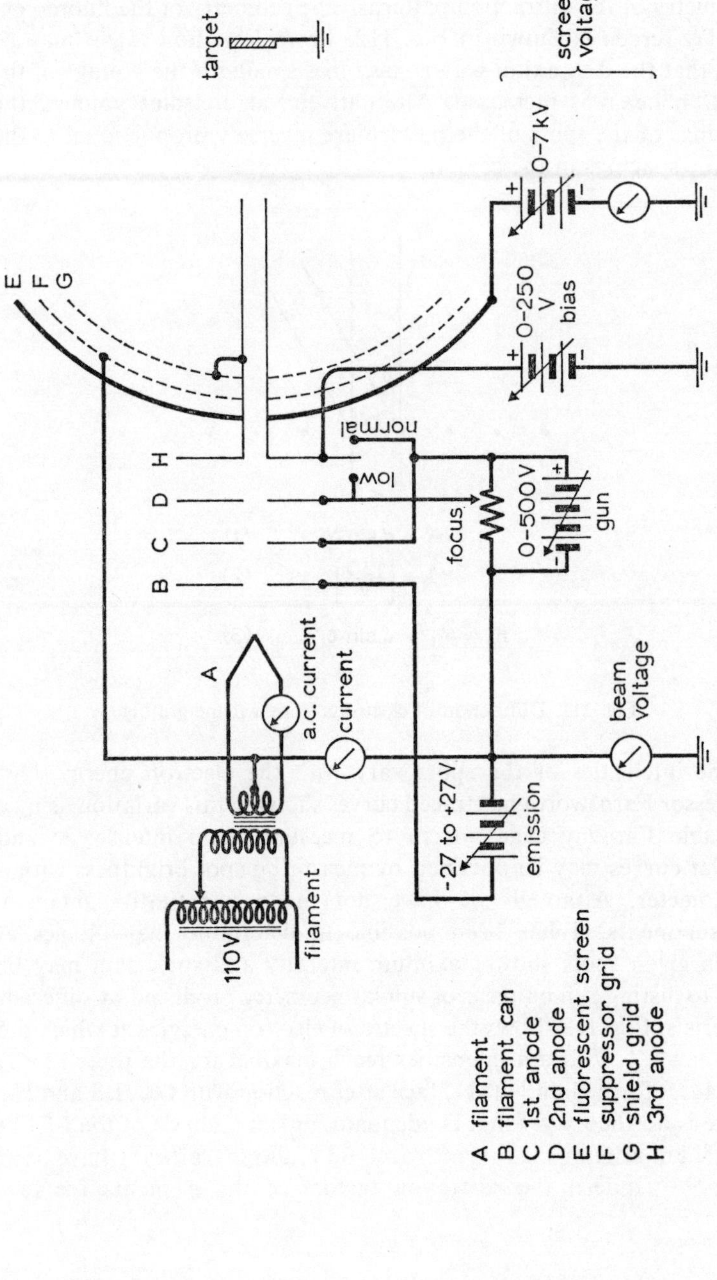

Fig. 110. The LEED electron optics (*Courtesy Varian Associates Ltd.*)

for the wavelength of the electron (λ is in Ångstroms when the electron energy V is given in volts) serves for the analysis of the geometry and symmetry of the diffraction patterns. The geometry of the fluorescent display screen is shown in Fig. 112. From Equation (5), it may be seen that the diffraction pattern becomes smaller if the voltage of the electron beam is increased. Alternatively, at constant voltage, the spacings of the spots of the pattern are inversely proportional to the atomic spacings in the crystal.

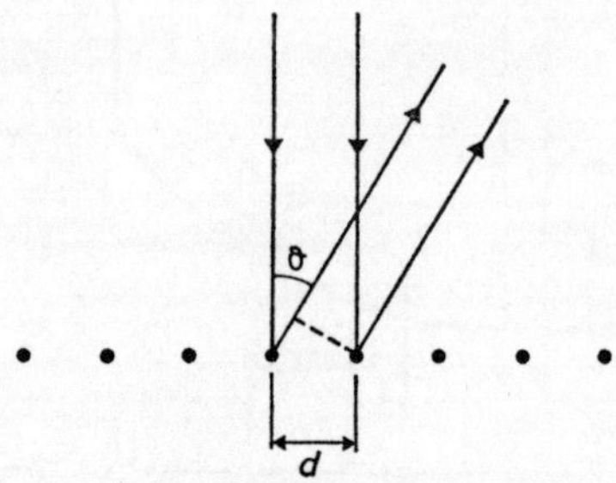

$$n\lambda = d \sin \theta \qquad (1)$$

$$\lambda = \left[\frac{150}{V}\right]^{\frac{1}{2}} \qquad (2)$$

$$n\left[\frac{150}{V}\right]^{\frac{1}{2}} = d \sin \theta \qquad (3)$$

Fig. 111. Diffraction of electrons from a plane grating

The intensities of the spots vary with the electron energy. [24,25] Professor Farnsworth has traced curves showing this variation using a movable Faraday cage system to measure beam intensity,[26] and similar curves may be obtained by measuring spot brightness with a photometer. Although we have not made quantitative intensity measurements, it has been possible to determine the voltages at which given spots show maximum intensity and such data may be used to distinguish patterns of similar geometry produced by different materials. Fig. 113 shows the spectra of electron energies at which the $(0, \frac{1}{2})$ and $(\frac{1}{2}, 0)$ beam intensities reach maxima for the three (2×2) structures formed on Ni(111) face after reaction with O_2, H_2S and H_2. There is no theory yet that is adequate for full analysis of the LEED data,[27] but attempts taking into account multiple scattering have been made.[28,29] Indeed the scattering factors of the elements for low

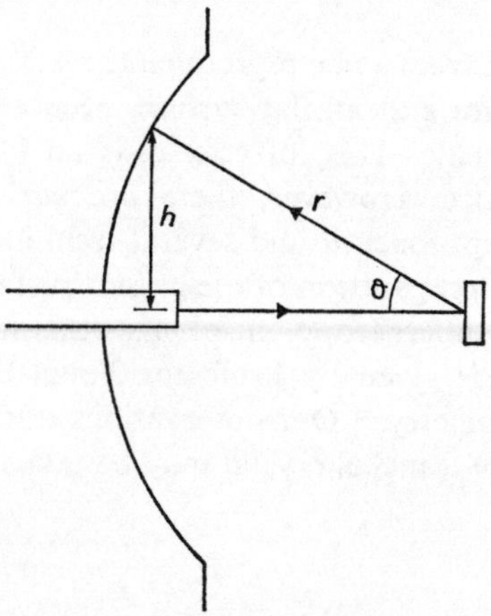

$$n\left[\frac{150}{V}\right]^{\frac{1}{2}} = d\sin\theta = \frac{dh}{r} \qquad (4)$$

$$hd = nr\left[\frac{150}{V}\right]^{\frac{1}{2}} \qquad (5)$$

Fig. 112. Geometry of fluorescent display screen

energy electrons are not well known.[30] It is not yet possible to deduce the surface structure from the diffraction patterns nor is it possible to calculate the intensity distribution over a pattern from a model structure, although the recent paper by Beeby[31] advances the technique for computing patterns. However LEED is an effective technique for examining ordered surfaces and their interactions with gases and vapours.

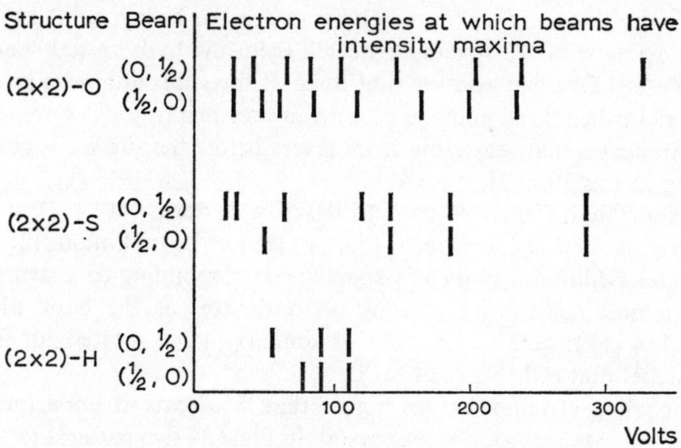

Fig. 113. Intensity maxima in electron beams diffracted from surface structures

Sample preparation

In such work it is necessary to start from some reproducible reference state. The aim is usually to prepare a clean flat surface, accurately orientated on a selected crystal plane. This, of course, is an ideal which is not attainable in practice. However, there are various methods available with which to approach it and several techniques which may be used to monitor the preparation of the crystal surface. We have used optical and electron microscopy but perhaps the most interesting picture of the surface is given by Professor Tolansky's method of multiple beam interferometry.[32] Our observations during the preparation of the (111) face of a nickel crystal may be taken as an example.

Ni(111)

The nickel crystal was a 6 mm diameter disc obtained from Metals Research, Cambridge, of 99·99 per cent purity, spark cut and polished with 1 micron and then $\frac{1}{4}$ micron diamond paste, and finally electro-polished in perchloric acid/acetic anhydride and thermally etched. Plate 32 is a multiple beam interferogram which shows the very smooth surface obtained in this way. Regularly spaced fringes indicate a surface of overall smoothness but the small irregularities indicate roughness of $\pm$20Å. The crystal was mounted and cleaned by bombardment with argon ions and annealed up to 450°C.

The (111) face of the face-centred cubic nickel crystal is hexagonal close-packed but it lies perpendicular to a three-fold axis. Plate 33 shows the diffraction pattern. One can see the hexagon of the first order spots with the spacing expected from the bulk crystal spacing of 2·49Å.[33] One can also see that three of the spots are more intense than the other three and the pattern in fact has trigonal symmetry. This indicates that scattering from layers below the surface is contributing to the diffraction pattern.

A shorthand system is used to describe patterns: (1×1) means the original first order pattern (Fig. 114 (a)) or its unit mesh; (2×2) indicates additional spots at $\frac{1}{2}$-spacings corresponding to a structure with a unit mesh twice as long on both axes as the basic nickel (Fig. 114 (b)); $(\sqrt{3} \times \sqrt{3})$R30° is similarly a cell rotated 30° from the nickel unit cell (Fig. 114 (c)).

Annealing at temperatures higher than 450° caused impurities to diffuse from the interior of the crystal. In Plate 34 two patterns (3×3) and (4×4) can be seen; the former appears first and is associated

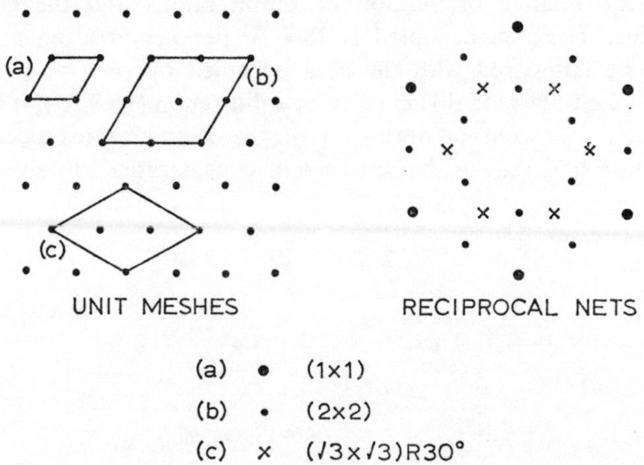

Fig. 114. Nomenclature of unit meshes and nets

with silicon, the latter with carbon. Prolonged treatment with oxygen at high temperature (830°C) followed by ion bombardment at 830° cleans the crystal to a greater depth and makes it stable at least for some hours up to that temperature.

The possibility of measuring rates of diffusion of trace impurities into the surface and observing the formation of the new surface structures is interesting in connection with the phenomena of adhesion of films and scale on metals. Compare this situation with that reported by Tylecote:[34] copper can be heated in dry air to form a cuprous oxide film which adheres when the sample is twisted. But if the metal contains 0·03–0·04 per cent phosphorus, the coating breaks away. This was attributed to the formation of cuprous phosphate but the details of the mechanism are certainly not well understood.

Low temperature sorption phenomena

A very significant investigation of the phenomena of physisorption and chemisorption on graphite was reported last year by Lander and Morrison.[35]

Fig. 115 shows the diffraction pattern of xenon physisorbed on graphite at 90°K. The additional spots due to the xenon monolayer correspond to a unit cell $(\sqrt{3} \times \sqrt{3})R30°$ and the lower diagram shows the relative orientation of xenon atoms and the graphite hexagons. The area occupied is 15·7 Å² per xenon atom and this should be compared with the area occupied by each atom in the close-packed hexagonal (111) face of solid xenon (16·8 Å²). It is clear that ca. 3 per cent compression of the xenon atoms occurs on adsorption and that in the presence of considerable adhesive forces

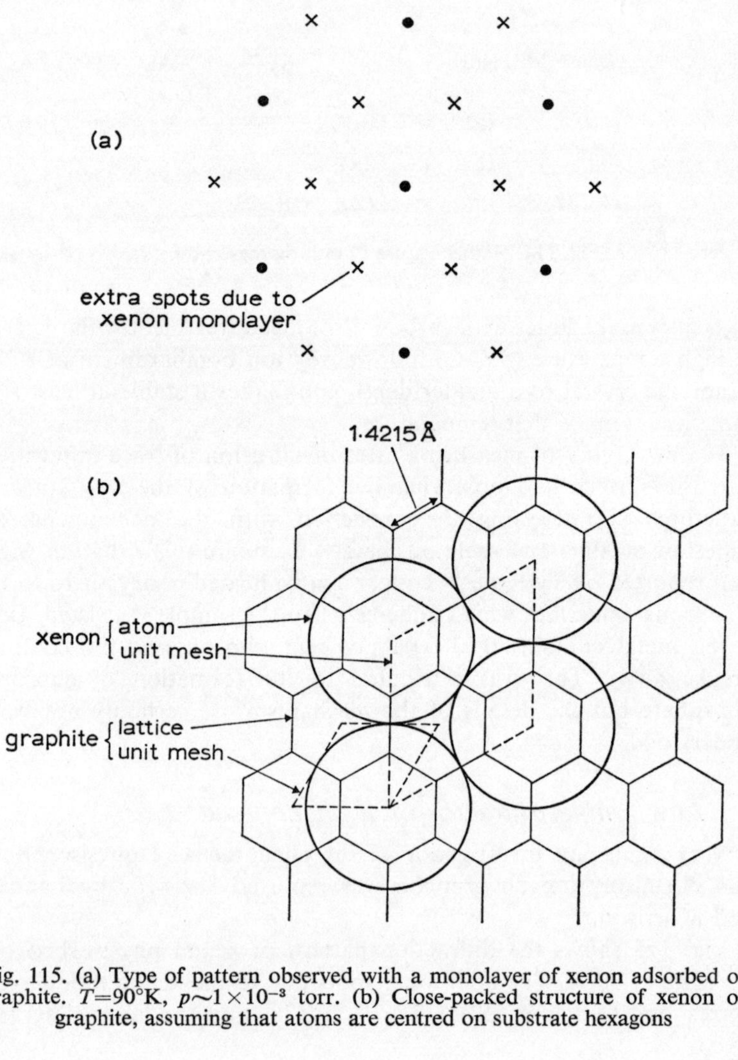

Fig. 115. (a) Type of pattern observed with a monolayer of xenon adsorbed on graphite. T=90°K, $p \sim 1 \times 10^{-3}$ torr. (b) Close-packed structure of xenon on graphite, assuming that atoms are centred on substrate hexagons

the strongly bonded solid surface is forcing a distortion on the adsorbed layer.

The importance of this phenomenon to the understanding of adsorption isotherms at $P/P_0 \geqslant 1$ and the distortion potential theory have been discussed by Adamson.[36]

Lander and Morrison also observed that adsorption of bromine on graphite occurred in several phases: gas, liquid, two distinct 2-D solid phases and multilayer phase, each stable in progressively lower temperature ranges. Bromine is relatively easy to observe with LEED but presumably other diatomic molecules, e.g. N_2, could behave similarly. Variations of experimental conditions could then have significant effects on surface area measurements; and the relevance of this to defining the condition of a surface at or during adhesion is evident.

The next examples show how chemisorption of sizeable organic molecules can be observed. Fig. 116 shows the diffraction pattern of sym-tribromo-benzene adsorbed on graphite at -10°C. The graphite

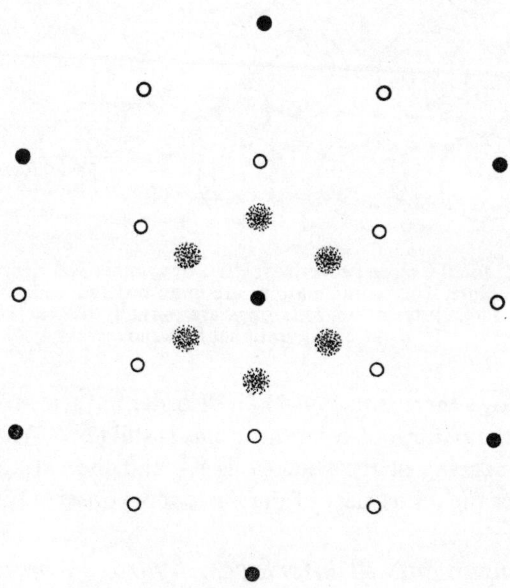

Fig. 116. Schematic of the pattern obtained from symmetric tribromobenzene adsorbed on graphite. The extra features are the sharp half-order spots and an anomalous set of fuzzy spots at about 8Å

pattern with 2·48Å spacing is seen and additional sharp spots at 5Å spacing in a (2 × 2) pattern. In addition there is a hexagon of fuzzy spots corresponding to an 8Å spacing.

The interpretation of this is that the bromine atoms register exactly with every second graphite ring giving the (2 × 2) unit cell with 5Å spacing. Although the molecules could have been arranged

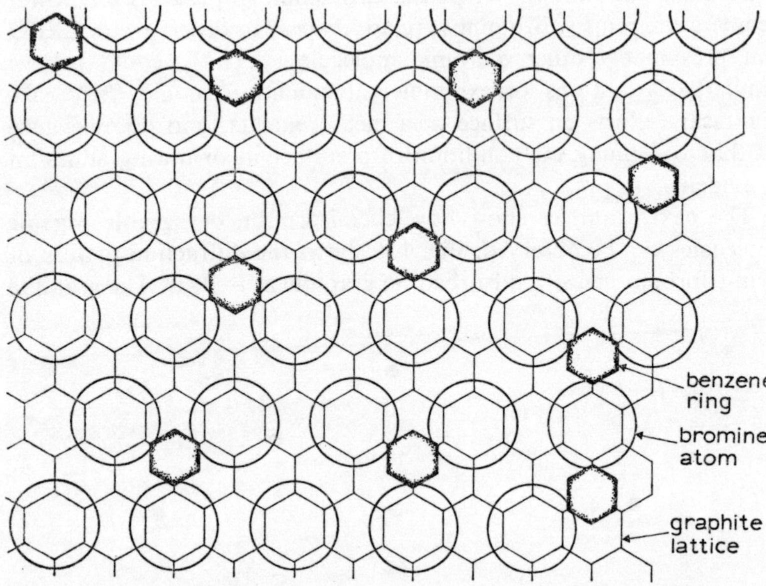

Fig. 117. Model for the close-packed structure of symmetrical tribromobenzene adsorbed on graphite. The bromine atoms are in an ordered array which is 2×2 with respect to the substrate. Benzene rings are partially disordered because of extra configurational freedom

regularly in rows there is in Fig. 117 a disorder in their arrangement; nevertheless the pattern of bromine atoms is still (2 × 2) and regular. The average spacing of the benzene is 8Å and the variation in this spacing causes the diffuseness of the inner spots observed in Fig. 116.

Rearrangements at interfaces: hydrogen on nickel

When hydrogen is admitted at 10^{-6} torr to a clean Ni(111) face additional weak and diffuse spots appear in the diffraction pattern (Plate 35) and reach their maximum intensity in about two minutes.

They become weaker on pumping to 10^{-10} torr at 25°C and disappear at 100°C. This (2 × 2) pattern resembles the pattern formed on the Ni(110) face[37] in that both correspond to meshes of twice the normal nickel spacings. The appearance of weak but well established and reproducible diffraction patterns attributable to weakly adsorbed hydrogen raises a number of questions. The intensity of scattering from hydrogen atoms is expected to be small compared with the metal atoms and diffraction must consequently arise from changes in the arrangement of the metal atoms. Germer and MacRae calculated from the shapes of the intensity versus electron energy (wavelength) curves of the Ni(110)–(2 × 1)–H beams, atomic spacings which correspond to nickel and deduced that the (2 × 1) structure was one in which the surface spacings of the nickel atoms was twice the normal value. This leads to the conclusion that adsorption of hydrogen on nickel promotes migration of nickel atoms from steps and defects on to the (111), (110) and possibly other faces in such a way that a new surface layer is formed with alternate rows of the underlying plane missing. On the (110) face the wider spacings occur in one direction only; on the (111) face the spacings are doubled in the three principal directions.

However, Ehrlich has reported[38] a suggestion made by G. W. Tucker that the same diffraction effects could be produced if adsorbed hydrogen caused displacements of adjacent rows toward and away from each other by even 10 per cent of the normal spacing. This model would also meet the requirements of the observed intensity variations and be more in line with the usual ideas of adsorbed layers. Ehrlich has pointed out that the agreement between various thermodynamic data does not support Germer and MacRae's long-range rearrangement model and concluded that the more classical picture of a slightly perturbed surface is still to be preferred.

Oxygen on nickel

The set of photographs, Plates 36–40 was obtained by Dr. T. Edmonds at Sunbury and demonstrate how new structures develop and change on the Ni(111) face with progressive treatment with oxygen at 10^{-7} torr. The patterns develop in turn, coexisting for a time with the previous one. The first oxygen induced pattern (2 × 2) (Plate 36) appears rapidly, reaches maximum intensity in 4–5 × 10^{-6} torr sec and then is gradually replaced by the ($\sqrt{3} \times \sqrt{3}$) pattern of Plate 37. The next stage (Plate 38) is a ($3\sqrt{3} \times 3\sqrt{3}$) structure which only appears if oxygen has previously diffused into the underlying layers of the crystal, and Plate 39 is the final NiO structure. Plate 40 is an example which shows a transition stage in which two structures coexist ($3\sqrt{3} \times 3\sqrt{3}$ + NiO). Fig. 118 summarises this sequence. The dot-dash arrow indicates that annealing at 250° destroys the oxide and restores the (2 × 2) structure. At 450° oxygen diffuses into a clean crystal and leaves a clean surface but in an oxygen-containing crystal heating to 850° may be necessary to remove oxygen rapidly

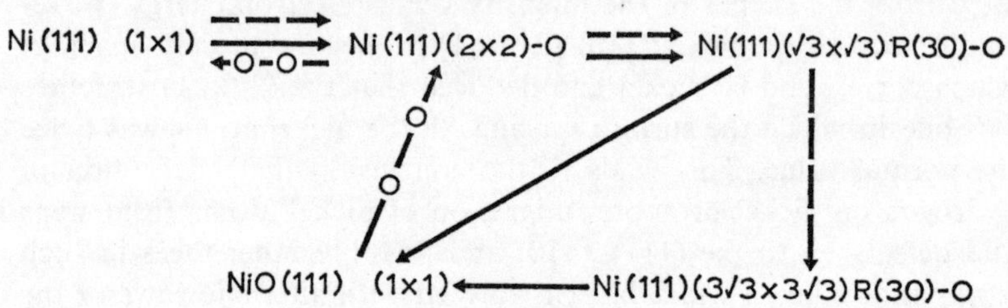

Fig. 118. Sequence of development of surface structure during reaction of oxygen with the nickel (111) face

from the crystal surface. The (2 × 2) structure is therefore quite stable and Park and Farnsworth have suggested that it is the surface manifestation of the Ni_3O structure.[39]

The nickel oxide patterns (Ni_3O and NiO) are in registry with the nickel pattern. They must form by epitaxial growth in patches several layers thick. The former develops rapidly but its pattern is visible with electron energies as high as 200 volts. If a single layer of NiO were formed, a (6 × 6) Moiré pattern would have been expected because of the difference in atomic spacings, but was never seen.

The ($\sqrt{3} \times \sqrt{3}$) structure is associated with a weakly chemisorbed state of atomic oxygen which desorbs at 100°C or under electron bombardment. An important and interesting conclusion is that different parts of the face must have different properties and presumably these arise from varying step concentrations, sub-surface structure, etc.

Similar complexities have been observed in extensive studies of the oxidation of copper.[40–42] Various faces react at different rates and a variety of 2-D oxide structures other than Cu_2O are formed. Such a detailed body of information, even though it is not completely understood or fully interpreted could make the basis of a programme for correlating surface pretreatment with strength of adhesion to metals.

A model for polymer adhesives

As a final example, we return to graphite and the adsorption of quite high molecular weight hydrocarbons. This work was prompted by the observations of Groszek[43] on the heats of adsorption of n-paraffins on lamellar solids, such as graphite and MoS_2. The isotherms of adsorption from solution in n-heptane showed sharp knees and virtual saturation beyond 2 g/l of n-dotriacontane. Examination in the electron microscope of oleophilic graphite which had high heat of adsorption (1·2 cal/g) for n-$C_{32}H_{66}$ but a low heat of adsorption of butanol (0·04 cal/g) showed that it consisted of large, very thin crystallites with a large ratio of basal plane surface to edges.

Breshchenko[44] observed some years ago that n-paraffins were strongly adsorbed on graphite and attributed this to localised adsorption arising from the matching of spacings in the paraffin chain with certain carbon spacings in the graphite. In fact, it is the H-H spacings between alternate methylene groups which match the graphite ring spacing. Molecules of long-chain n-paraffins can lie with minimal distortion in rows on the graphite face with a layer of H atoms lying exactly in the centres of the graphite rings in the energy minima.[45] The necessary distortions are: a contraction of less than 4° in the C-C-C angle and an increase of less than 2° in the H-C-H angle.[46]

If this were the mode of adsorption, accurate measurements of the area of exposed basal plane could be made. However, it seemed likely that a long segmented chain would adsorb in loops and not in this regular pattern,[47–49] unless nucleation took place at a step and the molecules had effective mobility presumably by repeated adsorption and desorption of segments.

Wheeler[46] attempted to resolve this problem using LEED. Graphite crystals extracted from Ticonderoga marble were prepared by cleaving and annealing at 500°. n-Eicosane was applied to the crystal from a simple molecular beam system but only disordered films which masked the graphite pattern were formed. Next, a graphite crystal attached to its holder was immersed in a solution of n-dotriacontane (1 g/l) in n-pentane to give monolayer coverage,[50] quickly rinsed with pentane, evacuated and examined with LEED.

At room temperature, weakened graphite patterns were seen and these disappeared on heating to 80°C.

Unfortunately the experiment does not give a clear-cut answer. Two of several possible explanations were (i) the adsorption was completely regular and the upper layer of H atoms was the source of the (1×1) pattern, weak because of a low scattering factor. The loss of pattern on heating could be attributed to disordering by thermal motion. Or (ii) the n-dotriacontane was deposited in small ill-formed crystallites or clumps and left areas of graphite uncovered. On heating above the melting point (70°C), the paraffin spread over the surface in a disoriented layer which masked the graphite structure. Further work is evidently required to resolve this problem and radioactively labelled hydrocarbons will be needed to confirm the coverages actually achieved on the crystals in vacuum. However, the possibility of handling and studying sorption of complex organic molecules in conditions suitable for LEED studies of ordered structures has been demonstrated and the modelling of adhesive systems is feasible.

Summary

It has been argued that difficulties met in the field of adhesion in planning basic research and even more important, using its results, have parallels in heterogeneous catalysis and other fields concerned with surfaces. These difficulties arise from the great complexity of practical systems and lack of understanding and control over the detailed structure of the surfaces and of the interactions of the solid surfaces. Examples are given to show the trend of recent advances in catalysis and particularly in the use of Low Energy Electron Diffraction.

LEED was shown to be a powerful technique for studying interphases, and the kinetics of the changes and reactions taking place in adsorbed films especially if used in conjunction with other techniques like mass spectrometry and radioactive tracers. Our own studies on materials from H_2 to a hydrocarbon with MW 450 have been carried out on systems selected for their relevance to catalysis and lubrication.

There is already a wide literature arising from studies on single crystal faces but particular reference has been made to the highly significant work of Lander and Morrison which clearly demonstrates distortion of atoms and molecules in sorbed layers. One can see

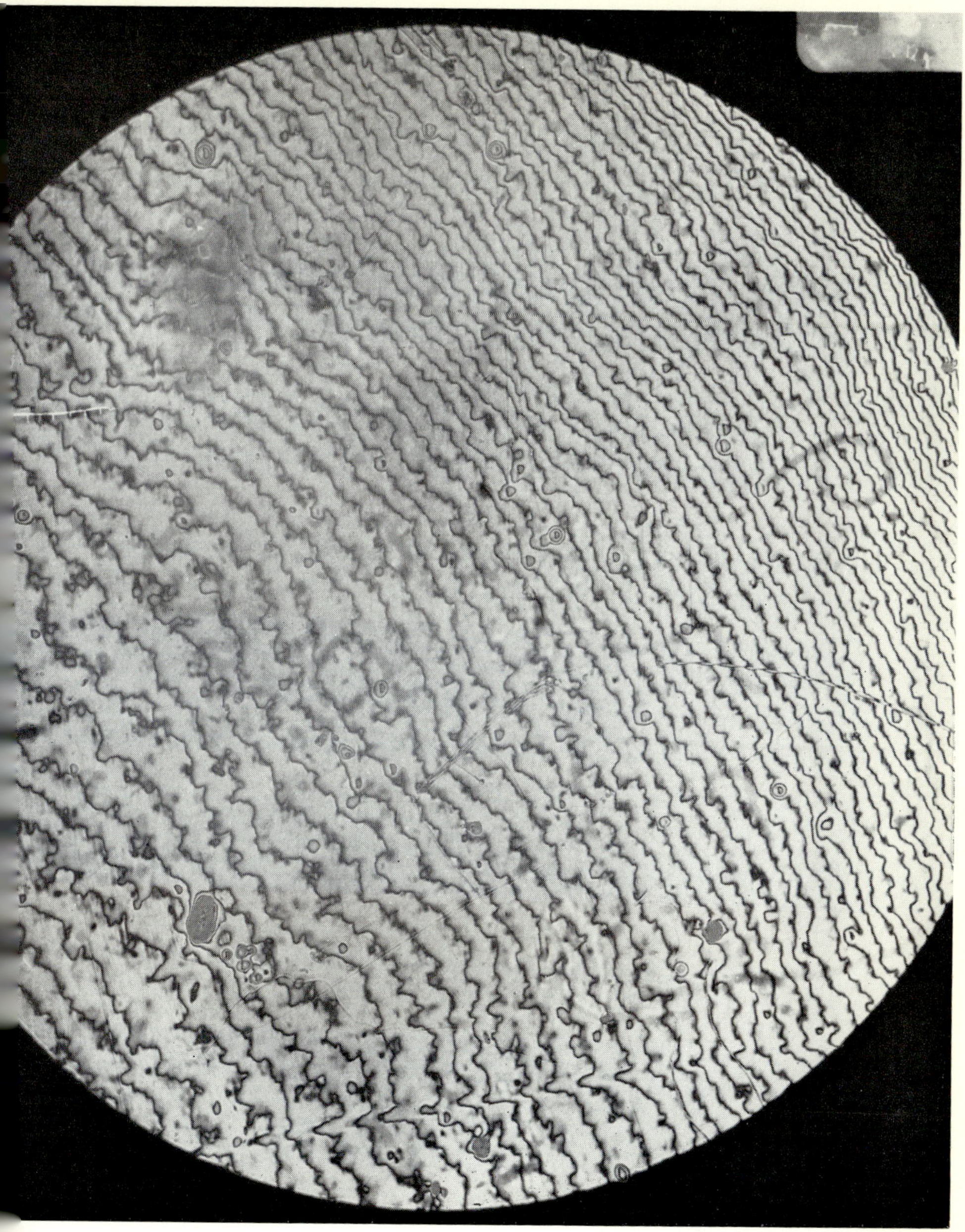

PLATE 32 Multiple beam interferogram of Ni (111) face after polishing

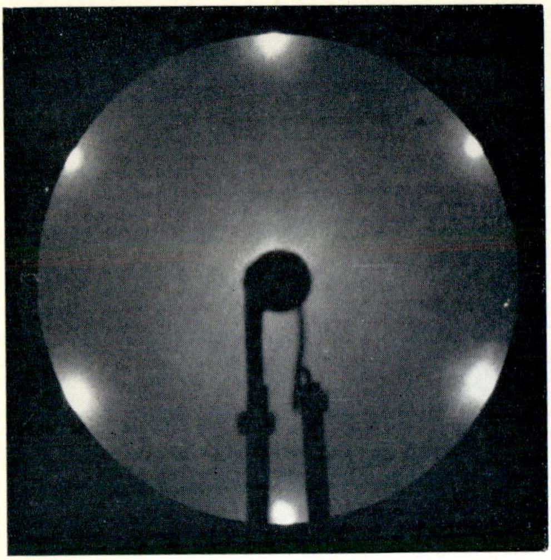

PLATE 33 First order diffraction pattern of Ni (111)

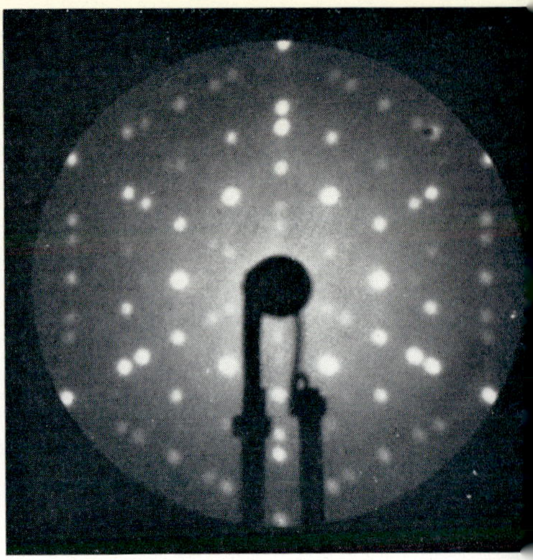

PLATE 34 (3×3)–Si and (4×4)–C diffraction patterns on Ni (111)

PLATE 35 Diffuse (2×2) diffraction pattern of hydrogen on Ni (111)

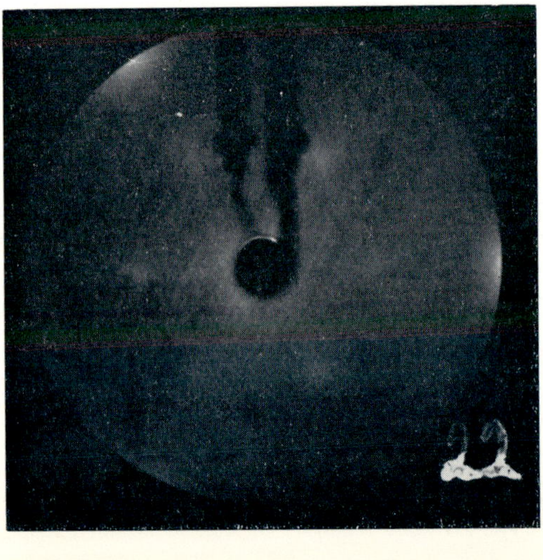

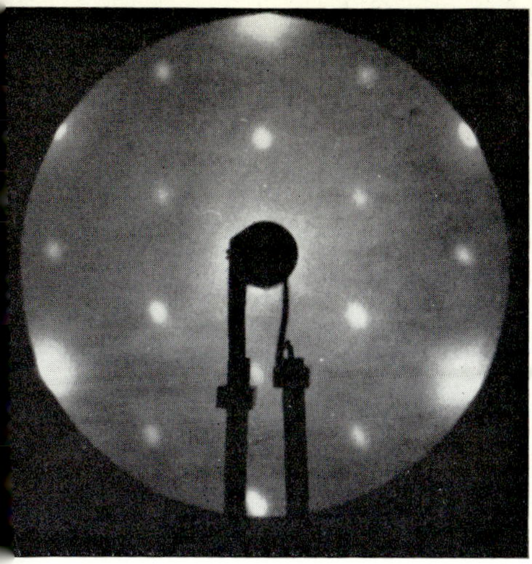

PLATE 36 Oxygen on nickel (111) first stage: (2×2)

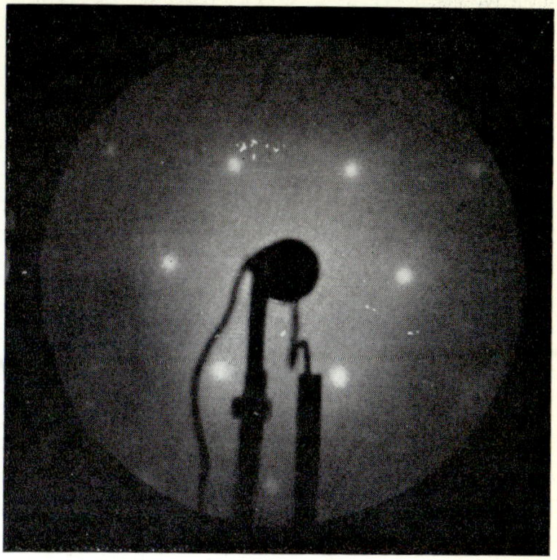

PLATE 37 Oxygen on nickel (111) second stage: $(\sqrt{3}\times\sqrt{3})R30^\circ$

PLATE 38 Oxygen on nickel (111) third stage: $(3\sqrt{3}\times 3\sqrt{3})R30^\circ$ appears only if oxygen has previously diffused into the crystal

PLATE 39 Oxygen on nickel (111) fourth stage: NiO

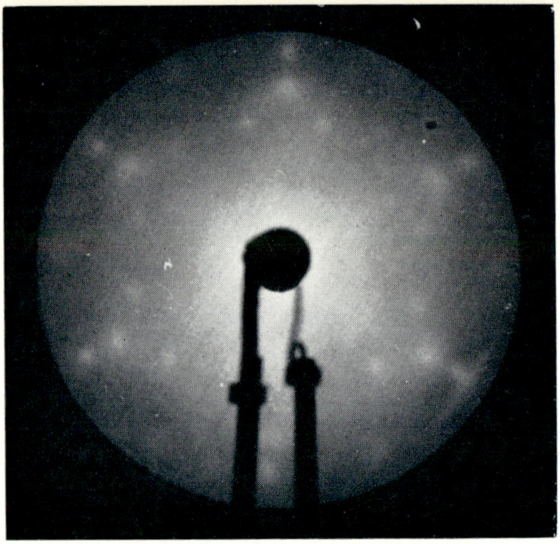

PLATE 40 Oxygen on nickel (111): coexistence of two structures $(3\sqrt{3}\times3\sqrt{3})R30°$ and NiO

parallel situations in adhesion problems and the great potential value of these developments for future research.

Acknowledgements

The author thanks the British Petroleum Company Limited for permission to publish this paper and gratefully acknowledges the permission to reproduce Figs. 115, 116 and 117, given by Dr. J. J. Lander and the Editor of *Surface Science.*

References

1. Anon., *Chem. Week*, Oct. 28, p. 117, 1967.
2. Mears, D. E. and Hansford, R. C., *J. Catalysis*, **9**, 125 and references given, 1967.
3. Kraft, M. and Spindler, H., *Proc. Fourth Int. Congress on Catalysis, Moscow, 1968*, Paper 69.
4. Tomezsko, E. S. J. and Furukawa, G. T., *J. Catalysis*, **9**, 386, 1967.
5. Eley, D. D., Moran, D. M. and Rochester, C. H., *Trans. Far. Soc.*, **64**, 2168, 1968.
6. Halpern, J., *Proc. Third Int. Congress on Catalysis, Amsterdam, 1964*, p. 146, Elsevier, 1965.
7. Cossee, P., *J. Catalysis*, **3**, 80, 1964; Arlman, E. J., *J. Catalysis*, **3**, 89, 1964; Arlman, E. J. and Cossee, P., *J. Catalysis*, **3**, 99, 1964.
8. Basila, M. R., Kantner, T. R. and Rhee, R. H., *J. Phys. Chem.*, **68**, 319, 1964.
9. Uytterhoven, J. B., Christner, L. G. and Hall, W. K., *J. Phys. Chem.*, **69**, 2117, 1965.
10. Ward, J. W., *J. Catalysis*, **9**, 225, 396, 1967; 10, 34, 1968.
11. Boreskov, G. K. and Karnaukhov, A. P., *Zh. Fiz. Khim.*, **26**, 1814, 1952; Boreskov, G. K. et al., *Zh. Fiz. Khim.*, **30**, 2560, 2787, 1956.
12. Dorling, T. A. and Moss, R. L., *J. Catalysis*, **5**, 111, 1966.
13. Boudart, M., Aldag, A., Benson, J. E., Dougherty, N. A. and Harkins, C. G., *J. Catalysis*, **6**, 92, 1966.
14. Poltorak, O. M. and Boronin, V. S., *Russ. J. Phys. Chem.*, **39**, 781, 1329, 1965; **40**, 1436, 1966.
15. Pitkethly, R. C. and Goble, A. G., *Actes du Deuxième Congrès Internat. de Catalyse, Paris, 1960,* p. 1851, Editions Technip, Paris 1961.
16. Hardeveld, R. van and Montfoort, A. van, *Surface Science*, **4**, 396, 1966.
17. Hardeveld, R. van and Hartog, F., *Proc. Fourth Int. Congress on Catalysis, Moscow, 1968*, Paper 70.
18. Hall, J. W., Ph.D. Dissertation University of Texas 1963 (University Microfilms Inc. Ann Arbor, Michigan, 64–6601).
19. Jaeger, H., *J. Catalysis*, **9**, 237, 1967.
20. Bourne, K. H., Holmes, P. D. and Pitkethly, R. C., *Proc. Third Int. Congress on Catalysis, Amsterdam, 1964*, p. 1400, North-Holland Publishing Company, Amsterdam, 1965.
21. Mueller, E. W., *Science*, **149**, 591, 1965.
22. Ehrlich, G. and Hudda, F. G., *Phil. Mag.*, **8**, 1587, 1963.
23. May, J. W., *Ind. Eng. Chem.*, **57**, 18, 1965.

24. Germer, L. H. and MacRae, A. U., *Ann. N.Y., Acad. Sci.*, **101,** 627, 1963.
25. Park, R. L. and Farnsworth, H. E., *Surface Science*, **2,** 527, 1964.
26. Farnsworth, H. E. and Madden, H. H., *J. Appl. Phys.*, **32,** 1933, 1961.
27. Gafner, G., *J. Appl. Phys.*, **2,** 534, 1964.
28. MacRae, E. G., *J. Chem. Phys.*, **45,** 3258, 1966.
29. Boudreaux, D. S. and Heine, V., *Surface Science*, **8,** 426, 1967.
30. Lander, J. J. and Morrison, J., *J. Appl. Phys.*, **34,** 3517, 1963.
31. Beeby, J. L., *Proc. Phys. Soc.*, **1,** (2), 82, 1968.
32a. Tolansky, S., *Multiple Beam Interferometry of Surfaces and Films*, Oxford Clarendon Press, 1948.
32b. Tolansky, S., *Surface Microtopography*, Longmans, London, 1960.
33. MacRae, A. U., *Surface Science*, **1,** 319, 1964.
34. Tylecote, R. F., *J. Inst. Metals*, **78,** 301, 1950.
35. Lander, J. J. and Morrison, J., *Surface Science*, **6,** 1, 1967.
36. Adamson, A. W., *J. Chem. Educ.*, **44,** (12), 711, 1967.
37. Germer, L. H. and MacRae, A. U., *J. Chem. Phys.*, **37,** 1382, 1962.
38. Ehrlich, G., *Ann. Rev. Phys. Chem.*, **17,** 37, 1966.
39. Park, R. L. and Farnsworth, H. E., *Appl. Phys. Letters*, **3,** 167, 1963.
40. Lee, R. N. and Farnsworth, H. E., *Surface Science*, **3,** 461, 1965.
41. Ertl, G., *Surface Science*, **6,** 208, 1967.
42. Simons, G. W., Mitchell, D. F. and Lawless, K. R., *Surface Science*, **8,** 130, 1967.
43. Groszek, A. J., *Nature*, **204,** 680, 1964.
44. Breshchenko, E. M., *Khim. i Tekhn., Topliv. i Masel.*, **9,** 32, 1957.
45. Poshkus, D. P., *Trans. Far. Sci.*, **59,** 177, 1963.
46. Wheeler, W. J., M.Sc. Dissertation, Brighton Tech. College, 1967.
47. Simha, R., Frisch, H. L. and Eirich, F. R., *J. Phys. Chem.*, **57,** 584, 1953.
48. Frisch, H. L. and Simha, R., *J. Phys. Chem.*, **58,** 507, 1954.
49. Frisch, H. L., *J. Phys. Chem.*, **59,** 633, 1955.
50. Davidson, E. and Kipling, J. J., A.C.S. Meeting, Atlantic City, September, 1965. Symposium on Colloid and Surface Chemistry in the Petroleum Industry.

The Effect of Surface Conditions when Bonding with Epoxy Adhesives

E. W. GARNISH, B.Sc., A.R.C.S. and C. G. HASKINS, B.Sc.

CIBA (A.R.L.) LIMITED

Introductory

Epoxy resins are important high strength adhesives. They are thermosetting materials presented in a variety of convenient physical forms, curing without pressure at temperatures ranging from about 4°C to 180°C. They adhere to many different sorts of substrate. The combination of good adhesive properties and versatility in processing makes them a natural choice in many bonding applications. However, practical experience has shown and adhesion theory can justify circumstances where epoxy resins exhibit less than exemplary adhesive properties. Polymer adherends with low surface energies and contaminated metal surfaces are cases in point.

This paper presents some investigations into this type of problem. The experiments described are of a technological nature but they do test and illustrate sundry aspects of adhesion theory. They also show that a system approach to adhesive bonding is essential. The adhesive is only one element in the system; the definition and treatment of the surface, factors in the hands of the practical user of adhesive, are equally important. One phase of the work was given to experiments with aluminium and steel adherends that had been submitted to pretreatments of varying rigour. The other phase was concerned with the bonding of polypropylene.

Experiments with metal adherends

There are three ways to join metals one to another; mechanical fastening, welding, adhesive bonding. Mechanical fastening, by its nature, is not dependent upon the state of the surface. The other two methods are. Any metal sheet, bar, etc., drawn from store is contaminated to some degree. The metal will have reacted with or adsorbed oxygen or other atmosphere gas, water and/or organic material. Also there may be massive protective coats of grease upon it.

For cleaning a soiled metal there are a range of techniques of increasing severity for removing the superficial and then the more difficult contaminants from the underlying rough microporous surface.

Upon most metals in air even when other materials are removed there will always be an oxide layer, but suitable pairs of metals will spontaneously adhere if the crystal beneath can be laid bare and touch. Such pairs have compatible crystal lattices and adhesion will be due to the strong metallic bond. This sort of bond will occur in certain kinds of welding where high energy techniques of pressure and/or temperature are used to displace the last contaminants and oxide layers.

One advantage of adhesive bonding is that it is used at modest temperatures and pressures that neither distort the metal nor affect the crystal structures adversely. In these conditions the bare metal lattice is never exposed. The most active coherent surface on which the adhesive may spread is a fresh, strong oxide layer with negligible or harmless adsorbed material. However, the cost of pretreatments to remove contaminants completely and leave such a layer can be an important element in a bonding process and so only those steps necessary and sufficient for the production of an adequate joint are tolerated. Since the type of adhesive and its curing conditions can assist in the displacement of surface contaminants, pretreatment and curing steps should be looked at in conjunction.

This was the complex of factors we studied with aluminium and steel lap joints. In the first set of experiments to be described a one-part epoxy paste adhesive was used. This adhesive has a threshold curing temperature of about 90°C. Four cure regimes were used varying from 2 hours at 95°C to 7 min at 180°C. There were two adherends; 20 s.w.g. mild steel to BS 15 and 16 s.w.g. alclad aluminium alloy to BS 2L73. Two levels of pretreatment were used upon the steel:

degrease in trichloroethylene vapour only, 10 sec.

degrease, shot blast with annular steel grit 2–4 sec, degrease again.

Three levels of pretreatment were used on the aluminium:

the two treatments used with the steel.

chromic acid pickling, after degreasing, to DTD 915B.

The latter is a wet process taking 30 min at 60°C, plus washing and drying steps.

To assess the adequacy of the joints, exposure conditions that were known to have an effect on the adhesive were chosen, viz. immersion in water at 80°C for seven days.

The results are summarised in the bar charts of Figs. 119 and 120, the Tables 32–35 and are given in full in the appendix. The bar charts show the general effects. The results on aluminium were better than those on steel. This was predictable since the former adherend was thicker and stiffer than the latter. The water immersion was the cause of falls in strength but the effects varied widely through the experiment. The results at the 80°C test temperature were at least as good as those at 22°C.

More detailed conclusions can be drawn from the summary tables. The initial tests in Table 32 show that shotblasting the steel gave better results than just degreasing, but the difference was not large; of the order of 10 per cent. The tests after immersion in water, however, gave the same ranking but the difference was now very much higher; of the order of 50 per cent. These are not unexpected conclusions. Other workers have demonstrated similar effects.

TABLE 32

Bonding Steel, the Effect of Pretreatment
Nominal breaking stress in lbf/in² quoted

	Initial test		Test after water immersion	
	22°C	80°C	22°C	80°C
Degrease	2 040	2 560	1 670	1 140
Shotblast	2 270	2 730	2 480	1 700

Means averaged over cures

However, Table 33 shows another effect; with this adhesive there was a dependence on curing regime. The initial tests indicated that the extreme curing temperatures give poorer results. The tests after immersion showed the same effect in a more extreme form; the joints cured at 180°C gave very poor results tested at 80°C after immersion.

Table 34 summarises the effects of pretreatment on the aluminium adherend. The initial tests showed the superiority of the well tried chromic acid treatment and the test after immersion emphasised this even more strongly. The experiment suggested that the structure of

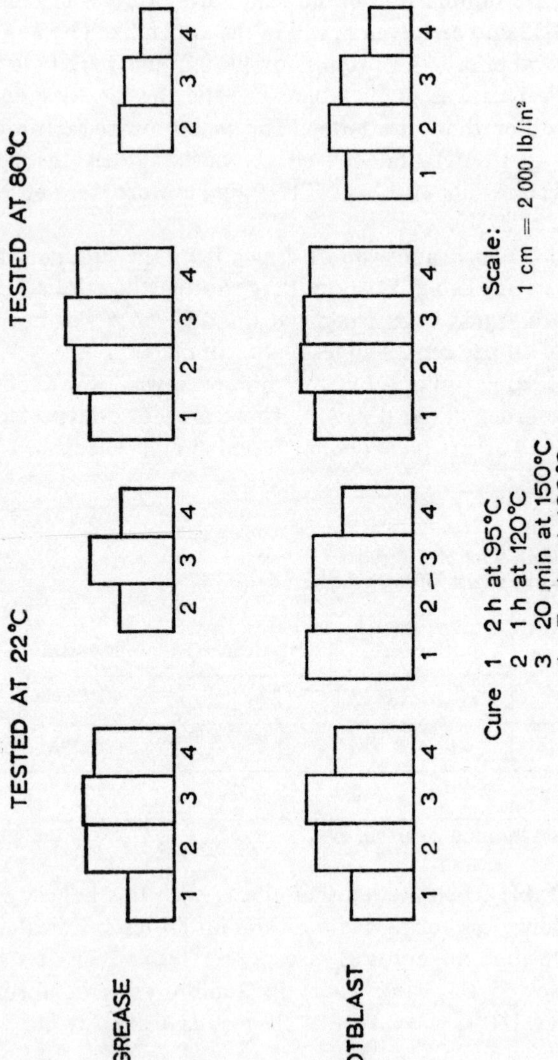

Fig. 119. Lap shear strength of paste adhesive on mild steel

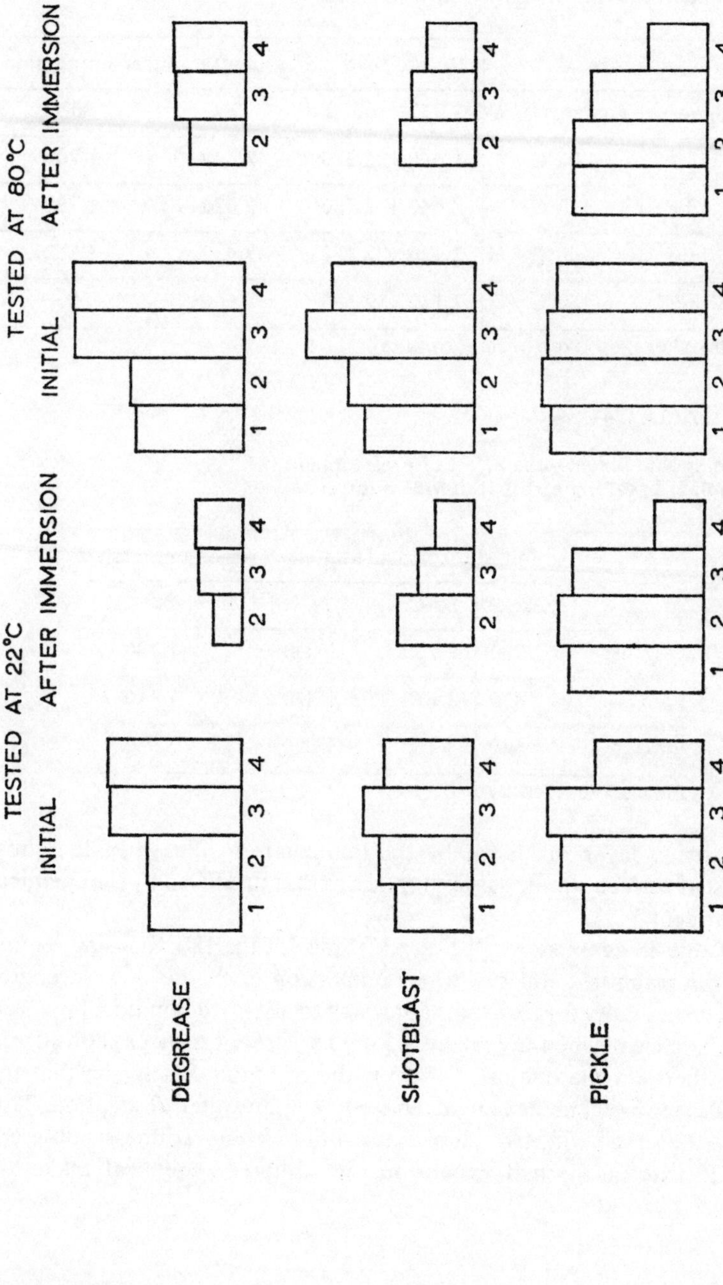

Fig. 120. Lap shear strength of paste adhesive on aluminium alloy

TABLE 33

Bonding Steel—the Effect of Cure
Nominal breaking stress in lbf/in² quoted

Cure		Initial tests		Tests after water immersion	
Time	Temperature	22°C	80°C	22°C	80°C
2 hr	95°C	1 440	2 370	2 760	1 700
1 hr	120°C	2 440	2 780	2 050	1 780
20 min	150°C	2 620	2 850	2 430	1 550
7 min	180°C	2 110	2 590	1 600	930

Means averaged over pretreatments

TABLE 34

Bonding Aluminium—the Effect of Pretreatment
Nominal breaking stress in lbf/in² quoted

	Initial tests		Tests after water immersion	
	22°C	80°C	22°C	80°C
Degrease	2 990	3 750	1 040	1 730
Shotblast	2 420	3 670	1 470	1 710
Pickle	3 670	4 150	3 090	2 960

Means averaged over cures

the oxide layer produced by the immediate oxidation of the shot-blasted surface was not as favourable as that produced by the chemical treatment.

Table 35 again shows the disadvantage of the 180°C curing regime by the magnified effect of water immersion upon the shear strength.

The necessity for a systems approach to adhesive bonding, however, is made clear when the results of the two sets of tests are considered together. On the one hand there is the common conclusion that the 180°C cure regime has an adverse effect upon water absorption. This was observed on two substrates of different stiffness, different coefficient of thermal expansion and different chemical make-up.

TABLE 35

Bonding Aluminium—the Effect of Cure
Nominal breaking stress in lbf/in^2 quoted

Cure		Initial tests		Tests after water immersion	
Time	Temperature	22°C	80°C	22°C	80°C
2 hr	95°C	2 250	3 330	3 630	3 530
1 hr	120°C	2 940	3 550	2 200	2 350
20 min	150°C	3 480	4 390	2 050	2 240
7 min	180°C	2 940	4 150	1 180	1 610

Means averaged over pretreatments

It is therefore a reasonable inference that it is an effect associated solely with the adhesive layer and would be observed with any other adherend that might be tried. On the other hand there are the differing effects of pretreatments. With aluminium resistance to hot water was poor when shotblasting was used; retention of the order of 50 per cent. With steel, however, the corresponding overall retention was of the order of 75 per cent. This level was reached upon aluminium only when chemical treatment was used. From these results we might make a general inference that the adhesive can retain three-quarters of its strength when joints are immersed in water at 80°C for seven days. However, the strong dependence due to substrate preparation gives the likelihood of an additional adverse effect. Thus the effect of water immersion with the optimum cure regime could not be predicted for another pretreated substrate from these results. The effect on the whole system, the joint, would again have to be measured.

It is of note with this hot curing epoxy system, that both degreased metals gave reasonable strengths in the initial tests. The pretreatment, dipping the metal in a trichloroethylene vapour bath, removes superficial grease but still leaves a layer of chemically bound contaminant on the surface. Water droplets placed on this surface show a high contact angle. The adhesive may not have disturbed this layer.

The adhesion that could be obtained with the simply degreased surface was explored further with a second type of epoxy system. This was a three part formulation mainly used as a coating, occasionally as an adhesive. The system is basically a modified aromatic amine liquid hardener and a liquid epoxy resin. With the aid of an accelerator the system will cure in ambient conditions. Whatever the temperature used there is only one crosslinking reaction taking place. The system is available in two forms; the first where an accelerator is a separate minor component, the second where a modified accelerator is built into one of two alternative hardeners. The first form has a mixing ratio of 100:60:1::resin:hardener:accelerator, the second has the ratio of 100:50–60:: resin:mixed hardeners. The second form gives wider possibilities in pot life by using easy proportions of the two hardeners.

These adhesives were tested on 20 s.w.g. mild steel. A good cure was obtainable with a regime of 24 hours at 22°C and 24 hours at 40°C but the joint strengths were found to vary widely with pretreatment. Results are shown in Table 36. Shotblast and acid etch pretreatments resulted in a mixed cohesive/adhesive failure in the joints; degreasing resulted in complete adhesion failure. The strengths showed the ranking: shotblast > acid etch ≫ degreasing.

TABLE 36

Cold Curing Amine-Epoxy system—Effect of Pretreatment
Nominal breaking stress in lbf/in² quoted

	Lap-shear strength		
	22°C	40°C	75°C
Degrease	660	630	610
Shotblast	2 130	2 120	1 200
Acid etch *	1 660	1 690	730

*Phosphoric acid in alcohol

Given the presence of a contaminating film on the degreased metal the adhesive to be effective would have to displace it or diffuse into it. With this system this was achieved by an elevated temperature cure. The effect is illustrated in Table 37. A burst of heat was applied to the joints before the standard cure. As the temperatures were increased the adhesive properties improved. There was a change from a weak adhesion failure to a strong mixed failure at the 70°C treatment level. The best treatment, ½ hour at 130°C, gave lap-shear strengths three to four times better than the control.

The heat treatments had been chosen arbitrarily so that there was doubt that the cross-linking reaction had proceeded to the same point. To check this the method was also tried on shotblasted steel. The best heat treatment, ½ hour at 130°C, improved the strength by only 50 per cent overall.

TABLE 37

Amine-Epoxy system—Effect of Burst of Heat
nominal breaking stress in lbf/in² quoted
All tests on degreased steel

Heat treatment before usual cure		Lap-shear strength		
Time (hr)	Temp. (°C)	22°C	40°C	75°C
24	40	870	730	650
2	70	1 570	1740	1 330
1·5	90	1 880	2 070	1 660
1	110	1 910	1 900	1 960
0·5	130	2 230	2 430	1 780
0·25	150	1 860	2 100	920

Each entry is the mean of at least eight tests

If the cross-linking reaction of the adhesive was allowed to proceed at room temperature the burst of heat became progressively less effective. Thus a burst of 15 min at 150°C after seven days cure at room temperature gave an improvement in strength but such a treatment made no impression after six weeks at room temperature.

The figures given in both Table 36 and Table 37 are mean values derived from two distinct sets of joints for each pretreatment. Although made at separate times the sets were not completely independent since they shared the same adhesive batch and metal consignment. The two hardener version of the adhesive came to be tested at a much later date and a different consignment of metal was used.

The tests on the 100:60:1 system were re-run as a check. Now it was found that the burst of heat necessary for good properties on the degreased metal was 1 hour at 110°C; the 70°C level was insufficient.

With the two-hardener version of the adhesive a 52:8 ratio of plain and accelerated hardeners gave good properties on the degreased metal only with the 30 min at 130°C regime. With a 30:20 ratio of hardeners all the regimes tested with the degreased metal were unsuccessful. The results were all low with adhesion failure.

Several conclusions can be drawn from these tests:

(i) if this adhesive be hot cured contaminated metal surfaces can be bonded;

(ii) a threshold curing temperature must be exceeded for good adhesive properties;

(iii) the threshold temperature is easily disturbed by small changes in the composition of the adhesive or in the conformation of the adherend or by advancement of the cross-linking reaction of the resin.

Our tests with adhesive systems, based on the modified amine and epoxy resin, could be viewed as tentative steps in the development of an adhesive process for bonding unpretreated steel. It is clear that the development is far from complete but it is also certain that a systems approach would be essential to bring the process to fruition. The composition of the adhesive, the necessary curing regime and the state of the metal surface; all would have to be more closely defined. The properties of the joints in service would have to be checked. The lack of reproducibility in the reported experiments represented a failure in control; the surface to be bonded had not been reproduced with the required degree of precision. However, even if our development were finished successfully the process might not be suitable for practical application because it lacked robustness. The balance between necessary substrate specification, adhesive definition and cure could be too delicate.

Successful adhesives for bonding oil covered steel exist and have established markets. Some of these were described at an earlier conference in this series.[1] Car manufacturers use these adhesives for secondary bonding in car body shells. The copious layer of oil on the steel sheet used assists the forming processes and is an indispensable protection against corrosion. It is not removed until just before the painting process. The adhesives give consistent results with a fairly high curing temperature and this feature, which is common to the adhesives tested here must be regarded as a necessary step in making these processes successful.

The idea that satisfactory bonds should be made on oil covered steel may seem unacceptable to adhesive technologists used to meticulous, carefully controlled cleaning and pickling pretreatment processes. The two concepts are reconciled, however, by a common principle that can be stated as 'in adhesive bonding the surface state of the adherend must be of controlled quality'. In the one case the specification of oil covering on the car makers steel is closely defined so that it can no longer be regarded as contaminant but has become an element in the adhesive process, just as in the conventional case the pretreatment steps too are an element in the adhesive process. Acceptance of this view does not imply equivalence between the two approaches. A thorough pretreatment process can be used successfully with a variety of adhesives and a range of curing conditions. The unpretreated surface has to be defined; some protective oils may be acceptable, others may not. The adhesive must be carefully chosen and it will require an elevated cure temperature.

The adhesives used with pretreated surfaces give higher strength joints than the adhesives that can at present be used with oily surfaces. Nevertheless the bonding processes that are worked with oily surfaces provide joints adequate for the purposes for which they are used. These processes have widened the market for adhesives and they will continue to receive the attention of the adhesive technologist.

Experiments with polypropylene

The tests made with polypropylene were all with commercial samples of polymer. We were attempting to make strong joints to a non-polar surface using polar epoxy adhesives curing in ambient conditions or at slightly elevated temperatures. There were several theoretical criteria to guide us in our experiments.

De Bruyne[2] has stated that to obtain strong adhesive bonds to a non-polar substrate it is necessary to use a non-polar adhesive. By this rule an unacceptable combination of materials was being essayed.

Polypropylene, because of its non-polar nature can only exhibit weak molecular interactions at an interface. However, even the weak dispersion forces that would interact with the polar adhesive predicate high adhesive strengths. For these interactions to be realised it is necessary for thermodynamic equilibrium to be reached and a large interfacial area to be attained on a molecular scale. According to Sharp and Schonhorn[3] this is possible in the adhesive joint if the liquid acting as the adhesive spontaneously spreads on the solid. Zisman[4] has shown that a liquid will spread on a low energy solid only if the surface tension is less than a critical value, γ_c. The critical surface tension may be determined by extrapolation of the plot of surface tensions of a series of non-spreading liquids versus cos θ, where θ is the contact angle against the solid surface. There is some variation in the parameter since a family of curves may be generated with different classes of liquids. For polyolefines γ_c is ca. 30 dyne/cm.

Also according to Zisman[5] the reversible work of adhesion between liquids and a low energy solid is given by the relation

$$W_A = (2 + b\gamma_c)\gamma_{LV} - b\gamma^2_{LV}$$

where b is the slope of the plot γ_{LV} versus cos θ. The maximum in this parabola for a particular surface can be determined. For polythene this occurs at $\gamma_{LV} = 54$ dyne/cm with $W_A = 76$ erg/cm^2. For a $-CH_3$ surface W_A max. is slightly lower but γ_{LV} is about the same.

Bikerman has shown[6] that adhesive properties will be strongly affected by stresses arising from volume changes during setting, by flaws at the interface and in the bulk phases, and by low-strength interfacial contaminants. The presence of the latter prevents a strong joint from being formed as was discussed in the previous section.

These were the four theoretical criteria which we could use in the problem. Our first step was to check the critical surface tension of the polypropylene with the type of liquid being used, viz epoxy resins. A commerical liquid epoxy resin (epoxy value 5·1 eq./kg) has a surface tension of 60–70 dyne/cm and gives a contact angle on polypropylene of 60–5°. Plasticisers are often used in epoxy resin adhesives, e.g. phthalate or phosphate esters. Such mixtures have lower surface tensions and contact angle. A Zisman plot made with mixtures of resin and plasticisers is shown in Fig. 121. The line was constructed by the method of least squares and the extrapolated value at cos $\theta = 1$ was 28·4 $\pm$ 9·7 dyne/cm (95 per cent confidence interval) in fair agreement with the literature values.

Now a simple mixture of an unmodified liquid epoxy resin with an amine hardener has a surface tension of 40–55 dyne/cm. By the Zisman equation this type of adhesive should be effective on polypropylene because the work of adhesion is at a maximum. We used double lap sandwich joints of the form shown in Fig. 122 to test the hypothesis. However, as many others have found the adhesion with

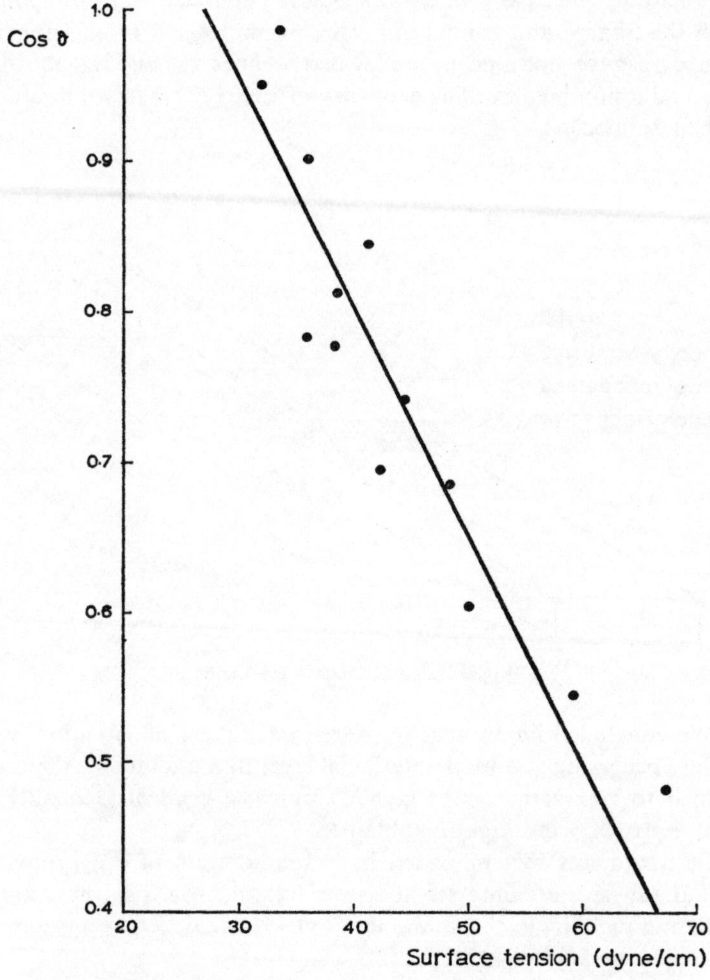

Fig. 121. Mixture of epoxy resin and plasticiser on polypropylene

such a simple room temperature cured adhesive was very poor.

We therefore chose from the plasticised resins one that had high wetting power with a surface tension of 36·7 dyne/cm and a contact angle against polypropylene of 26°. The addition of a modified aliphatic polyamine reduced this a little further although since the mixture gelled quickly precise figures could not be determined. We now had an adhesive with good spreading characteristics to comply with the Sharpe and Schonhorn criterion, but again room temperature cure gave poor properties with the broken coupons showing clean adhesion breaks. (The geometry of Fig. 122 was used in all the tests described).

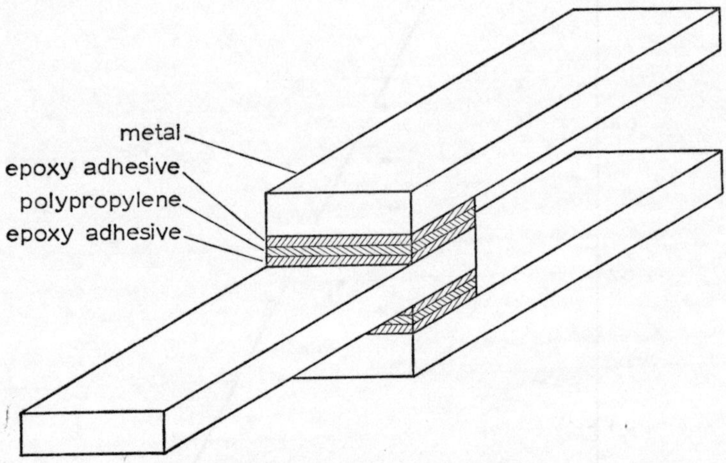

Fig. 122. Adhesive test specimens

We concluded finally that we were in a 'Bikerman situation' with failure occurring at a weak interfacial layer in the joint. We therefore turned to pretreatments to give a surface of controlled quality so that a proper joint might be obtained.

Pretreatments recommended by manufacturers of polypropylene to aid adhesive bonding are abrasion, treatment in a solvent vapour bath and pickling in chromic acid. We tried trichloroethylene vapour and chromic acid pickle.

A dip in trichloroethylene vapour gave a dull white finish to the smooth polymer surface. The effect of a light treatment on adhesion was spectacular. Joint strengths were such that failures of the polymer insert occurred.

The results are shown in Table 38. It was also clear that over-treatment could occur.

Trichloroethylene vapour was a severe physical treatment but chromic acid, in fact a mixture of alkali dichromate and sulphuric acid, was to be a severe chemical agent. The formula of the acid mix-

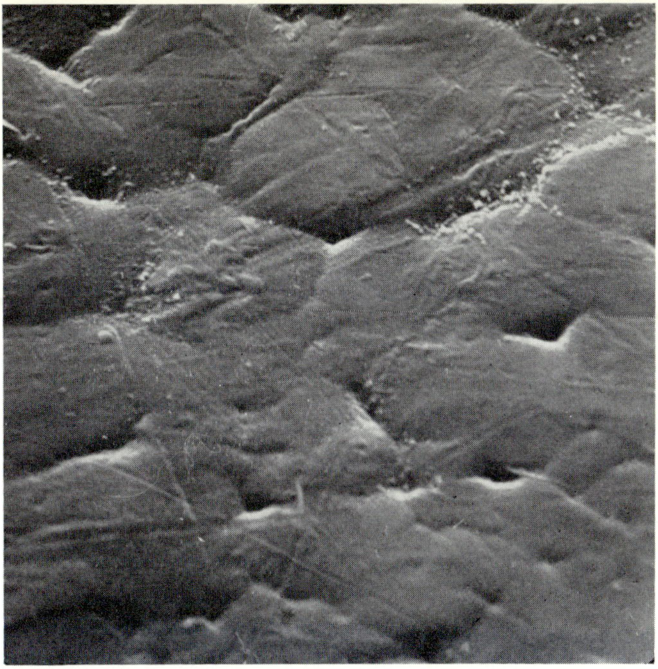

PLATE 41 Untreated polypropylene × 2500

PLATE 42 Untreated polypropylene from broken joint × 2450

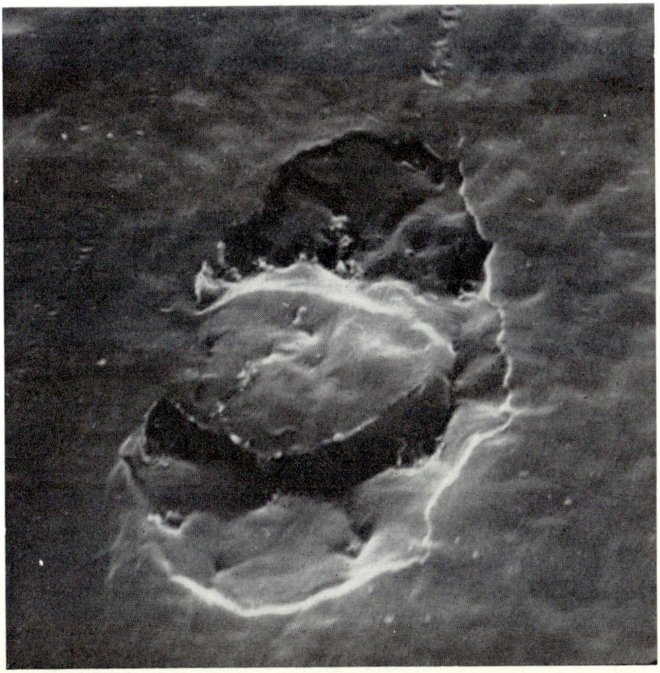

PLATE 43 Polypropylene exposed for 5 sec to trichloroethylene vapour × 2500

PLATE 44 Polypropylene treated with trichloroethylene vapour for 5 sec from broken joint × 2500

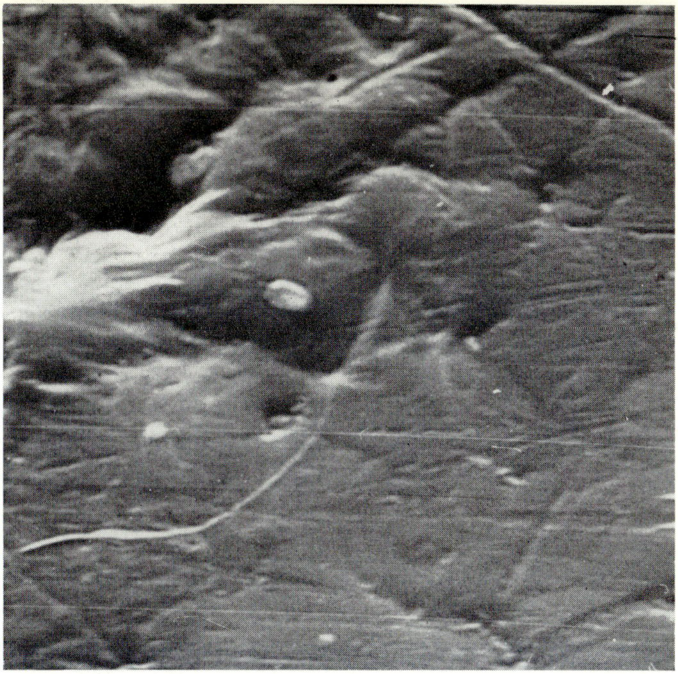

Plate 45 Polypropylene treated with chromic pickle for 15 min at ambient temperature × 2300

Plate 46 Polypropylene treated with chromic pickle for 15 min at ambient temperature from broken joint × 2500

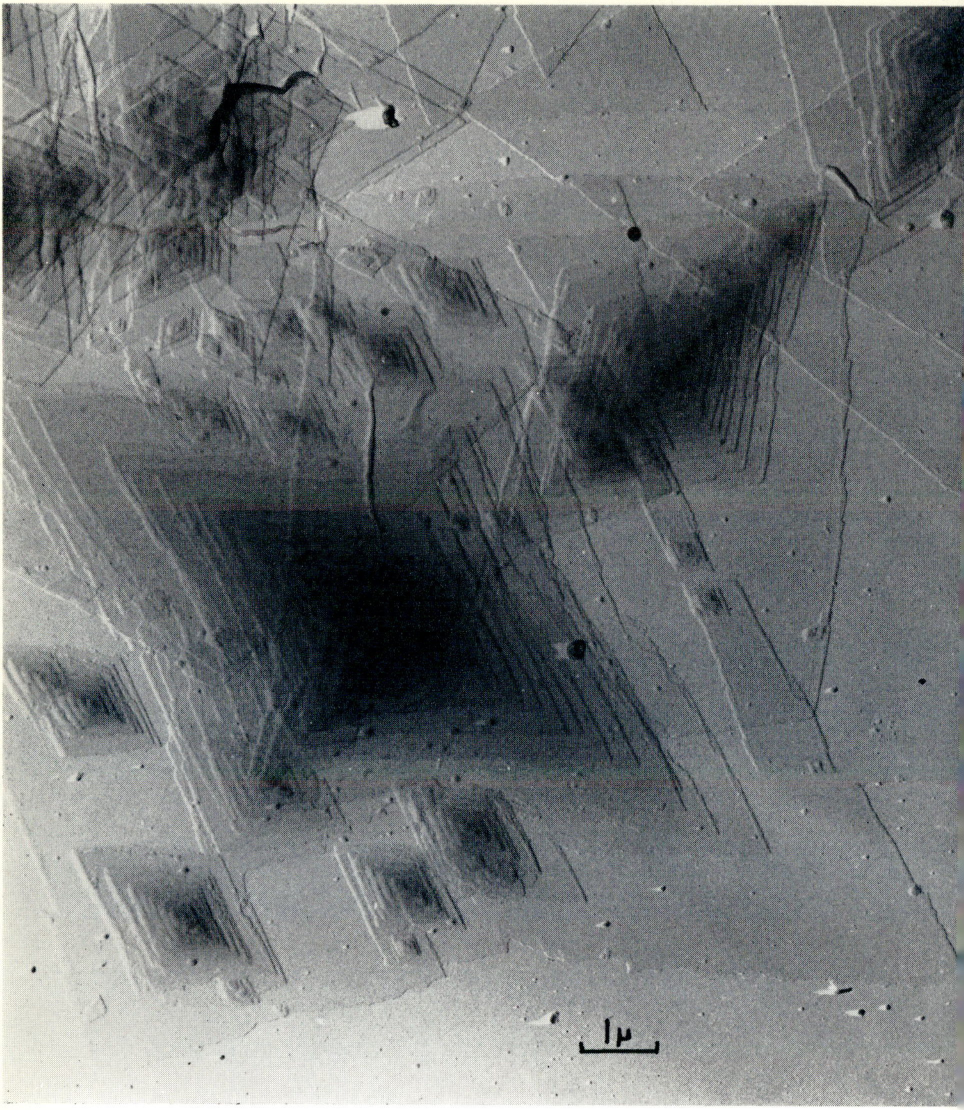

FLATE 47 Polyethylene crystal grown from 0·1% solution in xylene

TABLE 38

Effect of exposure to trichloroethylene vapour
Joints: Double-lap sandwich 1 in wide, ½ in overlap

Exposure (sec)	Joint strength (lbf)	Mode of failure
None	160	Adhesion
1	883	Polymer
5	859	Polymer
10	950	Polymer
15	300	Adhesion
30	133	Adhesion

Each entry is the mean of six tests

ture, the temperature and time of exposure could all be varied. We first submitted samples of polypropylene to the DTD 915B treatment and obtained joints that were as poor as those made with untreated polymer. We were successful, however, following an I.C.I. recommended formula (2·5 per cent w/v potassium dichromate in 31N sulphuric acid) used at room temperature. Results as shown in Table 39, were even higher than those obtained with trichloroethylene vapour.

TABLE 39

Chromic Acid Treatments

Treatment	Exposure	Joint strength (lbf)	Mode of failure
DTD 915B 7·5 pts $K_2Cr_2O_7$ 100 pts 6 *N*.H_2SO_4	30 min 60°C	168	Adhesion
2·5% w/v $K_2Cr_2O_7$ in conc. H_2SO_4	15 min 60°C	1 403	Polymer
	15 sec 22°C	1 193	Polymer
	1 min 22°C	983	Polymer
	2 min 22°C	880	Polymer
	5 min 22°C	1 207	Polymer
	15 min 22°C	1 190	Polymer
	30 min 22°C	1 127	Polymer

Each entry is the mean of six tests

In all the high strength specimens polymer failed by cracking, tearing or yielding. Complete cohesion failures were not observed.

TABLE 40

Pretreatment	Contact angle of	
	Water	Adhesive resin
None	90°	26°
trichloroethylene vapour	98°	34°
Acid pickle (I.C.I.)	40°	16°

The effect of the pretreatments upon wettability is shown by contact angle measurements on the polymer. Results are given in Table 40. The lower angle on the acid treated specimen shows chemical modification had occurred. The solvent dip surprisingly made the polymer less easily wetted, i.e. the ranking was:

vapour treated $<$ untreated $\ll$ acid treated

which may be compared with the adhesive strength ranking:

untreated $\ll$ vapour treated $<$ acid treated

Thus our experiment showed that wettability cannot be the sole criterion in determining the proper performance of a joint as a less wettable surface had given a much stronger joint.

The criteria of wettability was tested in another way by using room temperature curing epoxy adhesives that had not been formulated especially to spread on polypropylene. Results similar in mode of failure and strength to those of the high spread adhesive were obtained. The pretreatment of the polymer was therefore the determining factor.

The mode of failure was examined more closely with two adhesives using contact angles on polymer coupons extricated from the broken joints. Results and inference are given in Table 41.

The morphology of the surfaces, before and after bonding, was also examined using the electron scanning microscope*. The results are shown in Plates 41 to 46. The examination confirmed that the plain polymer had a weak covering film. This can be seen from Plates 41 and 42. The latter, a photograph of a debonded surface shows a section where the film had been torn. Elsewhere the surface

*Cambridge Instrument Company.

TABLE 41

Contact angles on broken coupons

System	Polymer treatment	Contact angle with water		Inference
		Polymer prior to bonding	Polymer from broken bond	
A	Untreated	77°	64°	Little change in surface
	Vapour dip	107°	105°	Little change in surface
	Acid treated	28°	72°	New surface produced
B	Untreated	77°	64°	Little change in surface
	Vapour dip	107°	90°	Little change in surface
	Acid treated	28°	96°	New surface produced

has the same appearance as the initial specimen in Plate 41. These photographs had to be taken very quickly since the electron beam attacked the weak layer as it scanned. The other plates, which are of pretreated polymer, did not suffer from the same difficulty since the surfaces were unaffected by the beam.

The trichloroethylene treatment caused the most drastic changes in surface morphology. A highly porous structure was formed which persisted in the broken coupon as well (Plates 43 and 44). This structure of lozenge shaped bodies did not correspond to spherulite nuclei. These were far larger and fewer in a cross-section of the polymer viewed by phase contrast microscopy.

The acid pickling process appeared to clean up the surface to reveal coherent polymer but without drastic effect on morphology (Plates 45 and 46). The contact angle had suggested that chemical modification to give some polar character had also occurred. The section from the corresponding broken coupon showed little change in appearance although the polar groups must have disappeared because the contact angle had increased. A new surface of the polymer or a very thin retained film of adhesive are possible alternative hypotheses.

The results given were all obtained with one grade of polymer. Extension of the tests to other grades is now showing that different samples show different susceptibilities to pretreatments. There is some evidence that this is linked with spherulite size in the samples so it appears that there is an added complexity in the moulding conditions to take into account when defining the adhesive process.

However, the experiments all show that control of surface quality is just as important when bonding polypropylene as when bonding metals.

Conclusion

In this paper we set out to explore certain aspects of adhesion as they might apply in the complex situation of practical bonding processes. With both polar and non-polar surfaces we showed that the properties of the adhesive joint are determined by an interaction of surface properties, adhesive properties and process variables. Bikerman's criterion that flaws at weak interfacial layer are causes of poor adhesive properties has been a useful hypothesis supported by the results of the tests. The experiments emphasise that a systems approach is always necessary to ensure valid conclusions when examining problems in adhesion.

References

1. Rubin, W., *Aspects of Adhesion 2*, p. 49, University of London Press, 1966.
2. De Bruyne, N. A., *Aircraft Eng.*, **18,** (12), 53, 1939.
3. Sharpe, L. H. and Schonhorn, H., *Advances in Chemistry Series*, (43), 192, 1964.
4. Zisman, W. A., *Ind. Eng. Chem.*, **55,** 19, 1963.
5. Zisman, W. A., *Advances in Chemistry Series*, (43), 38, 1964.
6. Bikerman, J. J., *Science of Adhesive Joints*, Academic Press, New York, 1961.

Appendix

Joint design: Single lap joint 1 in wide, 0·5 in overlap
Each entry is the main breaking load (lbf) of four specimens

Results with epoxy paste adhesive on 20 s.w.g. mild steel BS 15

Pretreatment	Cure cycle	Initial lap shear strength (lbf)				Lap shear strength after 7d immersion in water at 80°C (lbf)			
		22°C		80°C		22°C		80°C	
		Mean	Range	Mean	Range	Mean	Range	Mean	Range
Degrease	2 h 95°C	625	150	1 120	150	—	—	—	—
	1 h 120°C	1 143	320	1 323	310	717	120	695	200
	20 min 150°C	1 225	150	1 420	280	1 105	150	632	200
	7 min 180°C	1 080	220	1 253	250	690	120	388	200
Shotblast	2h 95°C	813	200	1 248	250	1 380	260	848	150
	1h 120°C	1 293	180	1 458	130	1 330	250	1 087	170
	20 min 150°C	1 400	200	1 415	200	1 327	110	922	110
	7 min 180°C	1 025	200	1 335	130	915	200	538	100

Results with epoxy paste adhesive on 16 s.w.g. alclad aluminium alloy BSL73

Pretreatment	Cure cycle	Initial lap shear strength (lbf)				Lap shear strength after 7d immersion in water at 80°C (lbf)			
		22°C		80°C		22°C		80°C	
		Mean	Range	Mean	Range	Mean	Range	Mean	Range
Degrease	2 h 95°C	1 213	450	1 450	800	—	—	—	—
	1 h 120°C	1 243	200	1 513	400	378	160	733	40
	20 min 150°C	1 730	500	2 250	100	588	50	925	300
	7 min 180°C	1 788	500	2 288	150	600	50	932	180
Shotblast	2 h 95°C	1 000	200	1 488	290	—	—	—	—
	1 h 120°C	1 238	100	1 668	350	975	450	1 013	50
	20 min 150°C	1 438	150	2 240	410	735	100	887	130
	7 min 180°C	1 175	150	1 938	500	518	150	662	300
Pickle	2 h 95°C	1 610	160	2 062	240	1 813	200	1 767	230
	1 h 120°C	1 930	250	2 140	280	1 945	760	1 775	350
	20 min 150°C	2 052	250	2 100	300	1 755	230	1 550	140
	7 min 180°C	1 440	390	1 985	90	655	50	822	250

The Morphology and Surface Energy of Polyethylene

M. J. RICHARDSON, B.Sc., Ph.D.
NATIONAL PHYSICAL LABORATORY, TEDDINGTON

In this paper attention is drawn to the serious discrepancies that exist between the surface energy of polyethylene as determined on the one hand by wettability measurements (say, 30–40 erg cm^{-2}) and on the other by the techniques (discussed below) of polymer physics (say, > 60 erg cm^{-2}). An attempt is made to resolve these, considering both the surface morphology and the nature of the surface energy measurements. It is only a slight exaggeration to state that until recently each group has been unaware of the work of the other. In the last year or so, however, Schonhorn has considered the differences and emphasised the need for a careful description of the morphology of the surfaces under consideration. This suggestion is probably the key to resolving the surface energy problem and present knowledge of the morphology of semi-crystalline polymers will first be considered.

Morphology

Until 1957 the behaviour of crystallisable polymers was thought to be explicable in terms of a 'fringed micelle' model in which different segments of a polymer chain are found in both crystalline and amorphous regions. A given chain traverses several crystallites which have dimensions of the order of hundreds of ångströms in the chain direction and microns laterally. Incorporation of chain segments in different randomly oriented crystallites effectively ties the remaining uninvolved segments with the result that later stages of crystallisation are hampered and extremely slow because they require much co-operative motion. Phenomena such as the growth of and, especially, the optical properties of spherulites are not really satisfactorily explained by this model and the discovery in 1957 of polyethylene 'single crystals' has resulted in many changes in our ideas of polymer morphology.

Single crystals

When a dilute (say 0·1 per cent) solution of polyethylene in xylene is cooled and the cloudy precipitate examined in the electron microscope it is found to consist, superficially at least, of perfect plate-like crystals (Plate 47). These are about 100 Å thick and may be several microns wide. Electron diffraction shows that the chains are oriented normally, or nearly so, to the large surfaces and therefore, since chains are of the order of thousands of ångströms in length, they must bend back on themselves to re-enter the crystal. This is beyond dispute but the re-entry mechanism, and hence the surface morphology, is controversial.

Two extreme cases may be envisaged—the regular fold (Fig. 123(a)) or the switchboard (Fig. 123(b)) model. In the former chains fold back on themselves in adjacent sites in a regular manner whereas in the latter they are permitted an excursion into the solvent before reincorporation in the crystal (we emphasise here that this is growth from a dilute solution so that a given molecule has little probability of being incorporated into two different crystals). In both cases the surface 'structure' is expected to be only a few ångströms in depth, and is not detectable in the electron microscope—the necessary shadowing material will be of greater depth and effectively conceal any surface detail on this scale. Great effort has been expended in finding unambiguous proof of which model is correct, so far in vain.

In general, microscopic methods favour the regular fold and macroscopic methods the switchboard. The former extrapolate the beha-

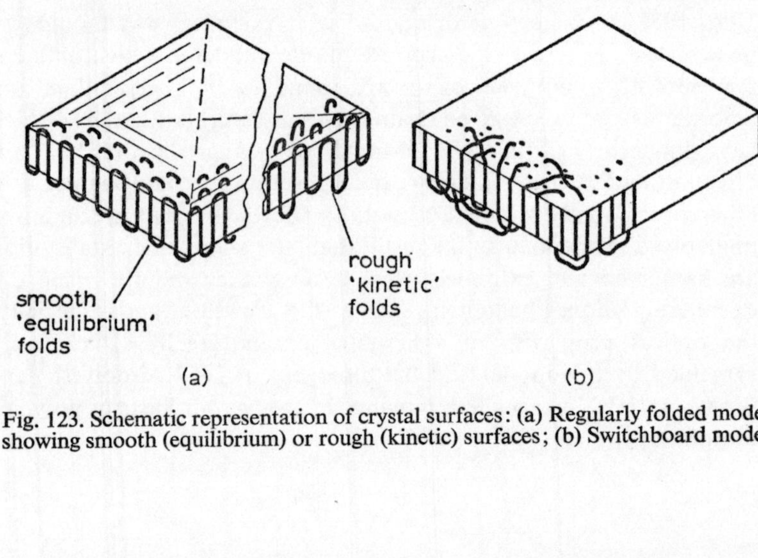

Fig. 123. Schematic representation of crystal surfaces: (a) Regularly folded model showing smooth (equilibrium) or rough (kinetic) surfaces; (b) Switchboard model

viour of individual crystals to the preparation as a whole whereas the latter are subject to the uncertainties due to solvent retention, the presence of holes and the imperfections of lamellar packing in general. So serious are these that even such apparently trivial properties as the density are in dispute.[1] It was thought that the surface energy of the crystals would provide at least a strong indication of the nature of the surface 'phase' and this is how we became interested in the problem at the National Physical Laboratory. Detailed discussion of the nature of the single crystal surface will be deferred until a later section. Here we make two comments pertinent to both solution and melt crystallised polyethylene.

First, the occurrence of thin polymer crystals (whatever the nature of their surface), so surprising initially, is, in retrospect, not only explicable but indeed essential. Thermodynamically an array of fully extended chains would be the stable configuration but for this to occur an 'infinite' nucleus is required and the probability of random fluctuations generating such a nucleus is normally minute (the deliberate introduction of foreign seed crystals is a special case and not considered here). However, entities of the order of a hundred ångströms in length and containing several 'chains' (which are in fact probably parts of the same chain) are constantly forming and redissolving in solution. With more supercooling these nuclei become progressively more stable and eventually grow rather than returning to solution. The initial size to a large extent determines the final thickness since large deviations from this are unstable in the sense that any protrusion involves additional surface energy. It should be stressed that the crystals themselves are metastable. To all intents and purposes they are stable below the growth temperature but thicken above this—whether by melting and recrystallisation or chain diffusion is not certain.

Our second comment concerns the search for a specific reactant for the surface and/or non-crystalline regions of polyethylene. Recently it has been shown[2] that fuming nitric acid appears to attack the surface phase of polyethylene single crystals much more rapidly than the crystalline bulk. There is an initial weight loss of some 10 per cent followed by only slight further losses with time. It is believed that this corresponds to the removal of the surface layer followed by a slow reaction with the remaining material. The molecular weight is drastically reduced showing that the re-entrant chains have indeed been attacked.

Melt crystallised polymer

To what extent are single crystals, grown from dilute solution, typical of bulk polymer, crystallised from the melt? Initially it was considered that single crystals were unique, interesting in themselves, but unrelated to the structure prevailing in the bulk. This view has now been modified since it appears that single crystals are indeed the basic structural units in melt crystalline material.

Surface replicas show that spherulites, the dominant morphological feature of bulk polyethylene, consist of arrays of ribbon-like fibrils. Digestion with nitric acid produces a brittle honeycomb-like material which powders under light finger pressure. The debris may be examined in the electron microscope and is found to consist of fragments remarkably similar to portions of solution grown crystals although rather thicker. This is to be expected. As the crystallisation temperature is increased (say from 70° to 90°C) the thickness of solution grown crystals also increases (100 to 150 Å); melt crystallisation invariably takes place at higher temperatures—commonly in the 120° to 130°C range. Remarks similar to those given above concerning the nucleation and growth of crystals from dilute solution are relevant also for melt crystallisation. In addition, since the growth medium is pure polymer, different segments of the same chain may become incorporated in two or more growing crystallites so that this feature of the original 'fringed micelle' model is retained.

The major difference is that in the more recent theories a crystallite surface is thought to contain many more re-entrant chains (that is, chains which already participate in the crystallite). Mechanical properties, for example, may then be discussed in terms of crystallographic slip and the effects of the 'tie' molecules that are involved in more than one crystallite.

A general theory of bulk polymer crystallisation has been given which accounts for the development of spherulitic structure from the original nucleus.[3] Bulk polymer is normally a mixture with a wide distribution of molecular weights. Crystallisation initially involves the highest molecular weight material. As growth proceeds low molecular weight 'impurities' are rejected and concentrate around the growing crystal. Further growth would be inhibited if this did not develop fibrils which project out beyond the impurity-rich region into virgin melt beyond. These fibrils, which are the structures containing re-entrant chains, therefore develop relatively rapidly radially. Low molecular weight impurities are concentrated between them, as far as the boundaries of the growing spherulite, where they remain in the amorphous state until perhaps the temperature is lowered, when they in turn may crystallise. Since, however, the interfibrillar-space is limited this secondary crystallisation will be slow as some material will be incorporated in more than one fibril and prevent the complete assimilation of all 'impurity' without difficult co-operative motion. Of course the rejected impurity may additionally contain material which cannot crystallise under any circumstances—branches or, for polymers other than polyethylene, stereoirregular forms.

Surface structure

The above brief survey of the morphology of polyethylene shows that it is possible, in principle, to conceive a whole spectrum of surfaces ranging from the totally amorphous (a spaghetti-like aggregate, devoid of any order) to the completely ordered (an array of extended chains, the regularly packed spaghetti rods before cooking). How is it possible to vary the surface morphology in practice and thus change the surface energy and adhesive properties? Conversely, what do surface energies tell us of the nature of the surface?

Here we consider the work of Schonhorn who, as mentioned earlier, appears to have been the first to suggest the importance of surface morphology. Starting with the observation that the usually accepted 'wettability' surface energy of polyethylene at room temperature (35 erg cm^{-2}) is virtually identical[4] with the extrapolated liquid (and hence amorphous) value[5] Schonhorn suggests that the surface of normal moulded polyethylene is indeed amorphous, being composed of rejected impurity as discussed above. He shows that the surface energy of single crystals is distinctly higher, say 50 erg cm^{-2}, the absolute value depending on certain assumptions.[6] Polyethylene formed by moulding against a high energy surface (gold) has an even higher surface energy, nearly 70 erg cm^{-2}.[7] In this last case the high energy surface must be dissolved away rather than stripped off as this also removes the unique high energy surface leaving the normal low energy surface behind. Certain measurements suggest that this unusual 'transcrystalline' surface has a very low amorphous content being composed of a large number of small crystals that can only develop at right angles to the gold surface.

Surface energy

At this stage we give an indication of the way in which the surface energy is determined using the methods of polymer physics and then return to Schonhorn's suggestions following the results of these and the above survey of polyethylene morphology. The comprehensive reviews by Hoffman[8] and Gornick and Hoffman[9] survey techniques other than our own. Here we indicate the methods and add some comments that seem relevant in the light of our own results.

Considering the regular fold model (Fig. 123) it is a simple matter to show that

$$\Delta G_L = \Delta G_\infty - (2\sigma_e/L) \quad . \quad . \quad . \quad . \quad . \quad (1)$$

where ΔG_L and ΔG_∞ are the free energies of fusion (per unit volume) of the crystals of thickness L and infinity (that is, the perfect, fully extended chain) respectively, and σ_e is the Gibbs surface energy of the large, chain folded face. Appropriate modifications may be made to permit the presence of amorphous surface layers (the switchboard model) but we do not give the details here.[10] The melting point, $T_{m,\,L}$ of these crystals follows as

$$T_{m,\,L} = T_{m,\,\infty}\left(1 - \frac{2\sigma_e}{\Delta H_\infty L}\right) \quad . \quad . \quad . \quad . \quad (2)$$

where $T_{m,\,\infty}$ and ΔH_∞ are respectively the melting point and heat of fusion (at $T_{m,\,\infty}$) of the extended chain crystals. This equation assumes that ΔH_∞ is independent of temperature in the range $T_{m,\,L} \rightarrow T_{m,\,\infty}$ but is otherwise exact. It is also valid for the lamellae in melt crystallised material. If σ_e is also independent of temperature a graph of $T_{m,\,L}$ against $(1/L)$ will give a straight line of intercept $T_{m,\,\infty}$ and slope $-2\sigma_e T_{m,\,\infty}/\Delta H_\infty$. This simple method has long been used to obtain both σ_e and $T_{m,\,\infty}$. Several points need comment, however. The melting point of polyethylene, single crystal (especially) or bulk, is not easily obtained; as noted above thickening occurs at high temperatures and to counter this extremely rapid heating rates are essential. Very often some kind of differential thermal analysis apparatus is used and thermal lag at the required heating rate has been ignored or corrected in a somewhat arbitrary fashion. We believe that both σ_e and ΔH_∞ are temperature dependent and also that ΔH_∞ is about 315 J g^{-1} rather than the commonly quoted 290 J g^{-1}. The variation with temperature is small but cumulative (σ_e decreasing and ΔH_∞ increasing with increase in temperature).

We have considered these points in some detail to emphasise that even with this apparently direct method there is some uncertainty in the required input parameters. Even when this is ignored and similar experimental techniques are used, very different surface energies are reported, values range from 57[11] to 127 erg cm^{-2}.[12] Some of the variation may be caused by real differences in surface morphology but much is attributable to the inherent experimental difficulties.

These, and uncertainties in the requisite constants, are even more important when kinetic methods are used to estimate the surface energy. These give $\sigma\sigma_e$ (from the radial growth rate of spherulites) or $\sigma^2\sigma_e$ (from homogeneous nucleation data) where σ is the lateral surface energy, that is, the free energy of a surface composed of crystalline CH_2-units. It is possible to fit most data originating from both single crystal and melt crystallised polyethylene and obtain consistent, temperature dependent, surface energies if certain assumptions with regard to $T_{m,\infty}$, ΔH_∞ and other quantities are made.[13] The 'kinetic' surface energy, $\sigma_{e(k)}$ (a distinction is made between a rough, as-grown 'kinetic' surface, and a smooth, annealed, 'equilibrium' surface), is given by

$$\sigma_{e(k)} = 68[1 + 0{\cdot}0135(145 - t)], \text{ erg cm}^{-2}, t\,^\circ\text{C} \quad . \quad . \quad (3)$$

The lateral surface energy is given by

$$\sigma = 13[1 + 0{\cdot}01(145 - t)] \quad . \quad . \quad . \quad . \quad . \quad (4)$$

Homogeneous nucleation studies on n-alkanes[14] have indicated that σ is some 10 erg cm^{-2} near room temperature so the two values are at least comparable. For polyethylene, melting and radial growth rate data are restricted to the range 120°–140°C, homogeneous nucleation occurs at about 90°C.

To summarise this brief discussion, therefore, we note that near these temperatures it is possible to fit much of the existing data from diverse sources with few assumptions. One of these concerns ΔH_∞ and we feel is incorrect; additionally the experimental melting point of extended chain crystals is only 141·4°C[15] so the suggested figure of 145°C may be too high—although here we readily concede that the experimental figure may be subject to upward revision when information becomes available regarding possible lattice imperfections. Our criticisms concern only absolute values of the surface energy and do not affect the general conclusion that, by these methods, the surface energy of polyethylene increases with decreasing temperature and even at high temperatures near 100°C is well above the room temperature 'wettability' value.

Current work

We consider next our own measurements of the surface energy of polyethylene. These were originally intended to determine ΔG_L

(Equation (1)) directly over a wide temperature range (−40°C to the melting point) for a series of polyethylene crystals grown at different temperatures and hence of different thicknesses. Using a precision adiabatic calorimeter, enthalpy increments and, normally, the corresponding entropy increment, over short (ca 10°C) temperature intervals have been determined. Cumulative graphs of the two quantities are of the form shown in Fig. 124. The molten portion may be extrapolated to low temperatures with some confidence for this turns out to be in excellent agreement with estimates obtained by extrapolating the properties of a series of molten n-alkanes to infinite chain length. True isothermal heats and entropies of fusion may thus be found (inset, Fig. 124) and hence the free energy of fusion (Fig. 125).

This is an oversimplified picture in that irreversible annealing affects the entropy, our apparent entropies of fusion are numerically

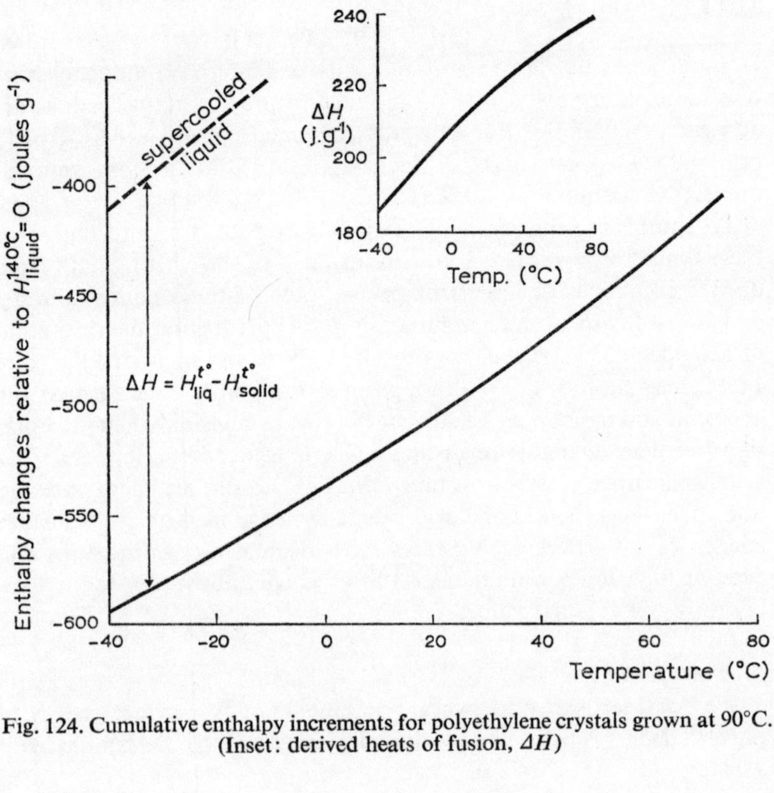

Fig. 124. Cumulative enthalpy increments for polyethylene crystals grown at 90°C. (Inset: derived heats of fusion, ΔH)

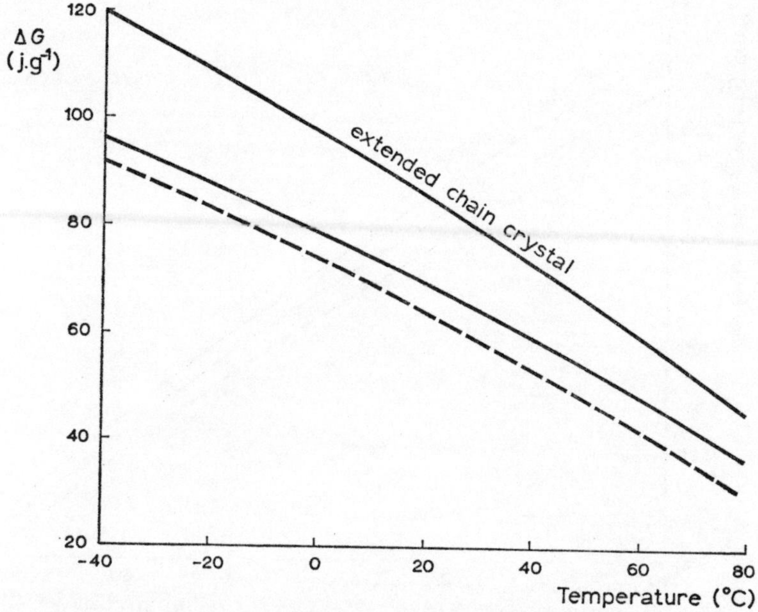

Fig. 125. Free energy of fushion, ΔG, of polyethylene crystals grown at 90°C (broken line, as measured; full line, corrected for irreversible entropy changes) and the perfect, fully extended chain crystals (top line)

too large and free energies of fusion too small. Fortunately it is possible, making only one very reasonable assumption, to estimate the requisite correction although a consequence is that the maximum temperature of our measurements is restricted to about 80°C with respect to the surface energy. The free energy of fusion of perfect, fully extended chains, ΔG_{∞}, may be estimated from those of n-alkanes of appropriate crystal structures, again over a wide temperature range. In this way only σ_e remains unknown in Equation (1) so here we have a fairly direct determination of σ_e and its variation with temperature in the range −40° to 80°C. The results are shown in Fig. 126. The differences between the crystals are small but probably real. Micrographs show that growth at 90°C gives relatively large and regular crystals (as in Plate 47), growth at 70°C gives much smaller irregular crystals. It would be surprising if the surface details were identical for these two extremes. The striking feature of Fig. 126 is the variation of σ_e with temperature. We feel that for these crystals this rules out the regularly folded chain model as such

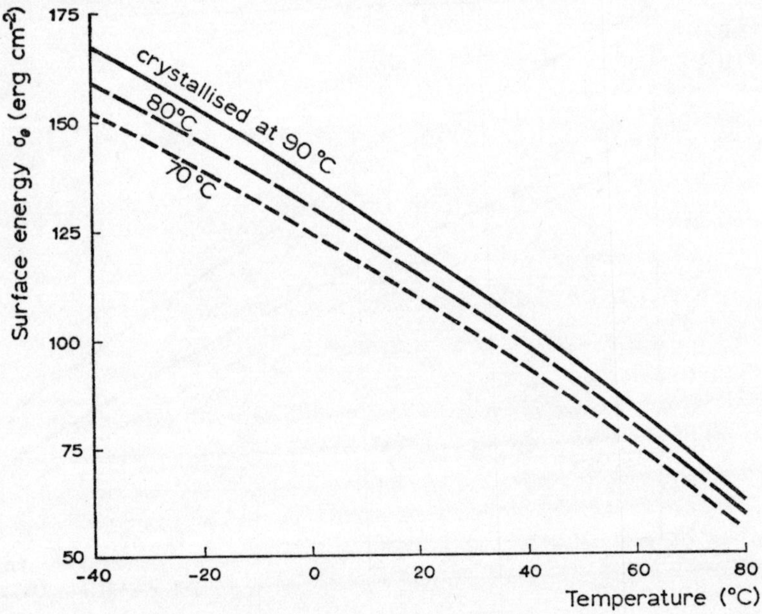

Fig. 126. Single crystal surface energies as a function of temperature. Crystallization temperature: 90°C (full line, L=145Å); 80°C (broken line, L=125Å); 70°C (dotted line, L=104Å)

(see below) but not the modified model, in which a slight variation in thickness is permitted resulting in a rougher, knobbled surface (the 'kinetic' surface referred to above). Recent theoretical work indicates that, of the two, the latter is indeed to be expected.[13]

Surface energy and morphology

Although the above results refer to solution grown crystals we recall that Equations (3) and (4) were obtained by fitting data from both these and melt crystallised material. The implication is that there are at least no *large* differences between the surface energies of the two types of polyethylene even though we have reservations as to some of the relevant input data and assumptions. What, then, do the surface energies of Fig. 126 tell us of the surface morphology?

As a prelude we note that for the idealised case of Fig. 123(a) σ_e should be given by[16]

$$\sigma_e = \sigma' + (q/2A) \quad . \quad . \quad . \quad . \quad . \quad . \quad (5)$$

where σ' is a surface energy similar to the lateral energy σ, q is the work to fold a chain back on itself and A the cross-sectional area of a chain. A value of some 70–80 erg cm^{-2} is thus indicated which is, most important, relatively insensitive to temperature. The observed temperature dependence is indicative of some kind of surface disorder and consequent appreciable surface entropy (the surface enthalpy is not expected to change by more than, say 20–30 per cent). Both the 'rough fold' and 'switchboard' models contain such entropy terms and it does not seem possible to decide on this basis alone which is correct. The case where an initially regularly folded model undergoes surface melting to give a looped amorphous surface layer can probably be rejected. We would expect this to happen at relatively high temperatures (certainly above room temperature) a constant 'regular fold' (Equation (5)) value being reached at the lower temperatures, this is contrary to observation.

It is very probable that, at least with melt crystallised polymer, something of both our two possible models is needed for a good approximation to a real surface, this being composed in the main of chains with adjacent re-entry, the folds giving a rough surface, but with numerous chain segments also incorporated in other lamellae, or non-adjacent parts of the same crystal. This is not very satisfactory in view of our wish to produce a morphological spectrum of differing adhesive properties. However, now that the basic morphological features are known, it is possible to consider routes whereby we can arrive at this goal. Starting with the regularly folded state this is probably already attainable at least on a microscopic scale. With care crystals can now be grown which, when suitably overlaying each other, show excellent Moiré patterns with clearly defined edge and screw dislocations.[17] It is virtually impossible to conceive this happening with anything other than very regular surfaces (which must still, of course, be folded) in intimate contact.

In addition Hoffman[13] has suggested that annealing normal solution grown crystals at temperatures below, say, 110°C (where gross thickening starts) is effective in reducing the surface roughness.

The work of Schonhorn has indicated a possible method of producing another type of crystalline surface. Crystals that grow in contact with high energy (specifically, evaporated gold) surfaces are soon constrained to propagate normal to the nucleating surface. The chains themselves are oriented parallel to the surface as sketched in Fig. 127(a). Evaporated gold is a polycrystalline material so the relative orientation of chains in different crystallites will be random, Fig. 127(b), even if there is some form of epitaxial growth on the gold nucleus—as seems to be the case when cleaved alkali halides are the nucleating surfaces.[18, 19] The many small 'monolayer' polymer crystallites soon meet and additional growth takes place normal to the high energy surface. On dissolution of the nucleation medium there remains, therefore, a randomly oriented array of extended (perhaps to the extent of a few hundred ångströms) crystalline —CH_2— units (that is, the lateral surface of polyethylene or n-alkane crystals). Grain boundaries exist that may correspond to those in the evaporated gold surface (if there is a high degree of

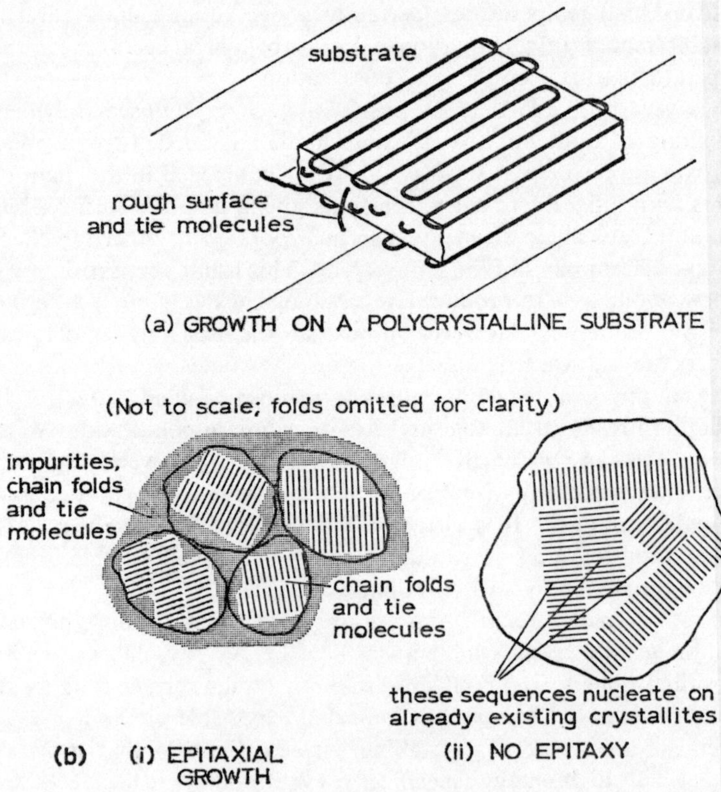

Fig. 127. Schematic diagram of nucleation on a high energy surface showing (a) the orientation of polymer chains, (b) random orientation of crystallites and grain boundaries

epitaxial growth) or may be on an even smaller scale if nucleation itself has little preferred orientation. The boundaries will be regions of chain folding, tie molecules, and general concentration of material unable to crystallise.

The work of Schonhorn indicates that the surface crystallinity is high so grain boundary effects may be small. The high surface energy, 70 erg cm^{-2},[7] is surprising, however, in view of the low 10 erg cm^{-2} from homogeneous nucleation studies[14] on essentially this surface configuration. If epitaxy occurs crystallites with a higher degree of orientation should be obtainable by growth from a single crystal substrate; fully extended chains may still not form as a growth nucleus of finite size would appear far more probable. In this respect we should mention the work of Wunderlich[15] on polyethylene crystallised under high (above 2 000 atm) pressures. This appears to give fully extended chain material, a most surprising development for which there is at present no satisfactory explanation. It would be of considerable interest to examine the surface energy of this material.

We now return to the surface energy of polyethylene moulded against perhaps a low energy polymeric surface or merely allowed to crystallise in an inert atmosphere. Normally we expect the surroundings to be at a lower temperature than the polymer so that crystallisation would commence *at the surface* (heterogeneous nucleation centres are invariably present—a fact which makes homogeneous nucleation studies so difficult) and proceed inwards, concentrating low molecular weight impurities in the bulk of the polymer rather than at the surface (the latter giving the amorphous layer suggested by Schonhorn). Micrographs of freely crystallised surfaces indeed show well defined crystals. Of course the opposite conditions could be obtained by deliberately initiating crystallisation in the bulk (incorporation cooling coils or a Peltier device) and here again the surface energy of polymer crystallised in the two ways would be of interest. Lacking this information how can we reconcile the 'wettability' surface energy with (i) an apparently crystalline (not amorphous) surface morphology, (ii) the high polymer physics values?

The answer to the first query is very likely provided in Equation (4), the free energy of the lateral surface—at room temperature this is about 30 erg cm^{-2}. Admittedly the result of a lengthy extrapolation, this does indicate that the crystalline surface may also have an energy akin to the 'wettability' values.

We feel that the second query above is invalid in that 'wettability' and 'polymer physics' surface energies are not really comparable. The latter measure the *total* surface energy which almost certainly includes a considerable amount of strain energy (chain folding) which is not specifically a surface phenomenon but happens to be included because of the particular morphology prevailing. 'Wettability' surface energies will not include this contribution. We would, therefore, expect the two types of measurement to give comparable values for surfaces where chain folding is absent (e.g. the lateral surface of planar zig-zag —CH_2— units). In this connection we should mention that measurements similar to those described above have been made on a series of n-alkanes of various crystal structure. For the orthorhombic and monoclinic forms the (terminal methyl) surface energy changes by only about 10 per cent over the range $-40°$ $\rightarrow 80°C$ around an average room temperature value of 60 erg cm^{-2}. Phillips and Riddiford[20] have reviewed the literature on this topic and conclude that the value is 33 erg cm^{-2}. Values quoted in Ref. 20 range from 20 to above 60 erg cm^{-2} depending on the technique and assumptions made. The highest figure, 63 erg cm^{-2}, is obtained by a 'zero creep' method[21] which Phillips and Riddiford believe is non-equilibrium but which in fact gives excellent results for the surface energy of metals. The remarkable agreement with our calorimetric value suggests a critical reappraisal of the assumptions inherent in the translation of results on sessile drops to solid surface energies.

Conclusion

In general we conclude that surface energy measurements on semi-crystalline polymers by sessile drop methods and the techniques of polymer physics should not give comparable answers for surfaces where chain folds are involved. Energies of other surfaces (e.g. fully extended —CH_2— units) should in principle be similar via the two groups but 'wettability' measurements may be subject to upward revision.

The morphology of polyethylene surfaces may be varied from amorphous to folded to extended chain with consequent differences in surface energy although we believe these to be slight. As far as adhesion is concerned chemical etching and/or discharge techniques, which introduce reactive sites, would appear to offer most opportunities for improvement in properties.

References

1. Martin, G. M. and Passaglia, E., *J. Res. Natl. Bur. Stds.*, **70A,** 221, 1966.
2. Blundell, D. J., Keller, A. and Connor, T. M., *J. Polymer Sci.*, A2, **5,** 991, 1967.
3. Keith, H. D. and Padden, F. J., *J. Appl. Phys.*, **34,** 2409, 1963.
4. Zisman, W. A., Contact Angle, Wettability, and Adhesion, *Advances in Chemistry Series*, Am. Chem. Soc., **43,** 1, 1964.
5. Schonhorn, H. and Sharpe, L. H., *J. Polymer Sci.*, **A,3**, 569, 1966.
6. Schonhorn, H. and Ryan, F. W., *J. Phys. Chem.*, **70**, 3811, 1966.
7. Schonhorn, H., *Polymer Letters*, **5,** 919, 1967.
8. Hoffman, J. D., *Soc. Plastics Eng. Trans.*, **4,** 1, 1964.
9. Gornick, F. and Hoffman, J. D., *Ind. Eng. Chem.*, **58,** 41, 1966.
10. Richardson, M. J., *Trans. Faraday Soc.*, **61,** 1876, 1965.
11. Brown, R. G. and Eby, R. K., *J. Appl. Phys.*, **35,** 1156, 1964.
12. Fujiwara, Y. and Yoshida, T., *Polymer Letters*, **1,** 675, 1963.
13. Hoffman, J. D., Lauritzen, J. I., Passaglia, E., Ross, G. S., Frolen, L. J. and Weeks, J. J., *Kolloid Z.* To be published.
14. Turnbull, D. and Cormia, R. L., *J. Chem. Phys.*, **34,** 820, 1961.
15. Arakawa, T. and Wunderlich, B., *J. Polymer Sci.*, **C,16,** 653, 1967.
16. Frank, F. C. and Tosi, M., *Proc. Roy. Soc.*, **A263,** 323, 1961.
17. Bassett, D. C., *Phil. Mag.*, **17,** 37, 1968.
18. Fischer, E. W., *Disc. Faraday Soc.*, **25,** 204, 1958.
19. Koutsky, J. A., Walton, A. G. and Baer, E., *Polymer Letters*, **5,** 185, 1967.
20. Phillips, M. C. and Riddiford, A. C., *J. Colloid Sci.*, **22,** 149, 1966.
21. Greenhill, E. B. and McDonald, S. R., *Nature*, **171,** 37, 1953.

Index

Author Index

Numbers in **bold** *refer to the volume*

Subject Index

Numbers in **bold** *refer to the volume*